DETROIT'S
LOST STOVE INDUSTRY

From Open Hearth to Cast Iron

Gerald Van Dusen

Published by The History Press
Charleston, SC
www.historypress.com

Copyright © 2025 by Gerald Van Dusen
All rights reserved

First published 2025

Manufactured in the United States

ISBN 9781467156998

Library of Congress Control Number: 2024947396

Notice: The information in this book is true and complete to the best of our knowledge. It is offered without guarantee on the part of the author or The History Press. The author and The History Press disclaim all liability in connection with the use of this book.

All rights reserved. No part of this book may be reproduced or transmitted in any form whatsoever without prior written permission from the publisher except in the case of brief quotations embodied in critical articles and reviews.

CONTENTS

Introduction 7

1. TWIN DRIVERS OF DETROIT'S FIRST INDUSTRIAL ERA 21
   The Discovery of Copper and Iron in the Upper Peninsula 21
   The Opening of the Soo Locks 26

2. VISIONARIES AND ENTREPRENEURS WHO BUILT DETROIT
IN THE LATE NINETEENTH CENTURY 29
   Eber Brock Ward: Detroit's First Millionaire Industrialist 31
   Dr. Samuel Duffield and the Origins of Parke-Davis 32
   Hazen S. Pingree: A Visionary Leader and Rooter
      of Corruption 34
   Mayor Christian Buhl and Detroit's Central Farmers' Market 37
   Captain Stephen Kirby and the Detroit Dry Dock Complex 38
   George Miller and the Beginning of Detroit's Thriving
      Tobacco Industry 41
   Berry Brothers Paints and Varnishes 42
   Dexter Mason Ferry and the D.M. Ferry Seed Company 44
   Bernard Stroh and the Stroh Brewery Company:
      A Detroit Beer Legacy 46
   Thomas Witherell Palmer and Elizabeth "Lizzie" Palmer:
      Shapers of a Unique Legacy 48
   James McMillan, John S. Newberry and the
      Michigan Car Company 51

Contents

3. From Open-Fire Hearth to Cast-Iron Stove 54
   Fire as a Primal Force 54
   The Birth of the Fireplace 55
   Early Experiments with Cast Iron: A Foundation
      for Innovation 57
   The Birth of the Cast-Iron Stove:
      Benjamin Franklin's Ingenious Contribution 58
   The Gradual Transition from Wood to Coal 59
   Why Coal? 62
   The Cast-Iron Stove Revolution 64

4. The Founding of Detroit Stove Works 65
   Jeremiah Dwyer: The Father of the Detroit Stove Industry 65
   The Detroit Stove Works: A Success Story 74
   The Process of Stove Making 76
   Detroit Stove Works Merges with Michigan Stove Company 79

5. The Michigan Stove Company 80
   George H. Barbour: Master Marketer 82
   The Giant Garland Stove at the Columbian National Exposition
      of 1893 85
   Exposition Universelle of 1900 87
   An Outrageous Tragedy 89

6. James Dwyer and the Peninsular Stove Company 93
   Who Was James Dwyer? 94
   The Labor Strike of 1887: Skilled Tradesmen Rebel Against
      the "Buck System" 96
   The Tragic Fire of 1893 97
   1909: Another Tragic Fire 99
   Workaday Hazards, Accidents and Injuries 102
   The Move to Brightmoor and Subsequent Bankruptcy 105

**7. Marketing Detroit's Cast-Iron Stoves in Detroit
and Around the World** 107
   Eliminating the Middleman 107
   Forward Integration 108
   Marketing with Trade Cards and Booklets 108

Contents

8. The Story of the Giant Garland Stove 113
   Design and Construction of the Big Stove 113
   Success in Chicago 115
   The Big Stove Returns Home to Detroit 117
   Move to New Home 118
   Big Stove, Big Move 120
   Into Storage at Detroit Historical Museum Warehouse 121
   Restoration and Return to the Michigan State Fair 122
   The Tragic End of the Big Stove 123

9. Commemorating Chief Pontiac's Rebellion:
The Michigan Stove Company's Bicentennial Tablet 126
   Historical Background 130
   Origins of Pontiac's Rebellion 130
   Tensions Escalate 132
   The Battle of Bloody Run 132
   The Disappearance of the Bloody Run Tablet 134

Conclusion 135
Bibliography 139
About the Author 143

INTRODUCTION

Now in its fourth century, the city of Detroit has proudly borne various titles, including Motor City, Arsenal of Democracy, Motown and Little Paris of the Midwest, and even suffered through a few unflattering nicknames. In the mid- to late nineteenth century, following its transformation from a French settlement to an agricultural oasis to a burgeoning manufacturing hub, Detroit would claim the title of "Stove Capital of the World."

During its peak, Detroit stood as the proud epicenter for the production and global distribution of more than 10 percent of all cast-iron stoves. The primary architects of this burgeoning stove industry in Detroit were Jeremiah and James Dwyer, who, both individually and in partnership, played a pivotal role in establishing the city as the foremost manufacturer of heating and cooking stoves. Their extraordinary journey from humble beginnings as orphaned siblings serves as a remarkable testament to their unwavering determination to leave an indelible mark on both the city and the industrialized world.

In the late nineteenth century, the Dwyer brothers were instrumental in founding and managing three out of the five major stove companies in Detroit: Detroit Stove Works, Michigan Stove Company and the Peninsular Stove Company. In today's monetary value, the Dwyer brothers' endeavors translated into the creation of a billion-dollar stove empire.

The narrative surrounding Detroit's flourishing stove industry carries profound historical, societal and cultural significance. The city's industrial involvement in facilitating the shift from open-fire hearths to stove systems represented a critical juncture in the evolution of household technology, fundamentally reshaping the methods by which people prepared meals and kept their homes warm. This transition was especially pronounced in the northern tier states, where a combination of climate factors and technological advancements played pivotal roles in driving this change. As time passed, the transition from open-fire hearths to more efficient and controllable stove systems yielded a multitude of advantages, not just impacting daily routines, but also reshaping social dynamics and cultural traditions.

However, despite the city's significant contributions to a global revolution in household technology, this and other flourishing Detroit industries—ships and railroad cars, pharmaceuticals, paint and varnish, tobacco and cigars, shoes and boots, packaged fruit and vegetable seeds, beer and others— would eventually fade into obscurity. As Charles K. Hyde asserts, "The first decade of the twentieth century represents a great divide in Detroit industrial history." The auto industry so dominates industrial history that the city's important industrial advances of the previous half century have been largely ignored, but many of these industries contributed significantly to the rapid and immense growth of the automobile industry.

Prior to Henry Ford's introduction of the moving assembly line on December 1, 1913, which reduced the time it took to build a car from twelve hours to ninety minutes, Detroit was already primed to become the center of the auto industry. Aside from its strategic geographical advantages, the city had established practically all the necessary industries vital for supporting auto manufacturing by the onset of the twentieth century. Along the riverfront, gasoline marine engines were being developed for personal transportation. By 1910, Detroit had more expert gasoline engine workers than any other city. It was not necessary to import or train a class of skilled tradesmen. The same could be said for carriage and wagon builders; factories could easily be retrofitted to produce motorized carriages. And there were many more (think: steel, copper, brass and non-ferrous metal foundries; product designers; carpenters, millwrights and mold-makers). In essence, Detroit had already established a robust infrastructure encompassing various industries crucial to support the burgeoning auto industry. This preexisting foundation set the stage for the rapid growth and dominance of automobile manufacturing in the region.

Regrettably, the narratives surrounding these industries have suffered significant neglect. Instead, scholarly and popular attention has predominantly gravitated toward the rise, peak and eventual decline of Detroit's automobile industry, shining a spotlight on the charismatic figures central to its narrative. Overlooked amid this focus is the intriguing tale of one such forgotten industry, Detroit's dynamic stove industry, emblematic of a time when the city boasted an impressively varied industrial landscape. This serves as a poignant reminder that Detroit once thrived with a remarkably diversified economy, producing commercial goods that achieved remarkable success and widespread recognition across the nation and throughout the world.

From Fur Trading to Family Farming

In the early nineteenth century, Detroit underwent a significant transformation, shifting from its reliance on fur trading as the dominant economic force to a more diversified economy, thanks to visionary regional planning and the discovery of abundant natural resources. This evolution paved the way for a multifaceted economic and social structure centered on agriculture, trade, commerce and manufacturing. Although it would take several decades to fully materialize, the initial stages of this transformation were quickly becoming evident.

Recognizing the unsustainability of fur trading as a long-term economic strategy, Michigan authorities were keenly aware of the vast untapped fertile land spread across the state, especially in southeastern Michigan, which was the first part of the state to be thoroughly surveyed. The potential for agricultural development was immense, provided that land acquisition costs remained reasonable and prospective settlers from the East could be persuaded to embark on the challenging journey to claim this promising land.

In 1818, the establishment of a land office in Detroit marked a crucial first step toward drawing in settlers from New York and the Eastern Seaboard to the Michigan Territory. At that time, the Michigan Territory had already been subdivided by early surveyors into expansive county regions, which were further divided into smaller township sections, each typically spanning thirty-six square miles. To entice prospective settlers, however, these vast geographical divisions needed to be transformed into marketable sections of eighty acres, in accordance with the federal government's minimum size requirement. Local entities such as George Jerome and Company dedicated significant efforts to furnish the land office with well-documented surveys, ultimately facilitating the sale of family-sized farms.

Introduction

Persuading easterners to make the trek west faced additional challenges, not the least of which were erroneous reports that Michigan Territory was uninhabitable. In 1814, General Duncan MacArthur, second in command at Fort Detroit during the War of 1812, wrote in a private letter, "From my observation, the Territory appears to be not worth defending, and merely a den for Indians and traitors. The banks of the Detroit River are handsome but nine-tenths of the land in the Territory is unfit for cultivation." Though the general's comments were privately communicated, the concern he expressed is likely to have been shared, word-of-mouth, by soldiers returning east and south after the war, possibly dampening some enthusiasm among those susceptible to such messaging.

Lewis Cass (1782–1866). *Library of Congress.*

In 1816, Edward Tiffin, surveyor general of the United States, reported to President Madison that Michigan was not worth further surveying because it was so inhospitable to settlers and not well suited for agriculture. Tiffin was relying on reports he had commissioned from surveyors sent north into southeast Michigan from Defiance, Ohio. The federal government had set aside two million acres of land as a reward for veterans of the War of 1812. These surveys helped to determine where veterans might be able to profitably settle. Unfortunately, the surveyors drew broad and unfortunate conclusions from time spent in the low-lying Monroe, Michigan area, which has had historical issues with flooding and drainage. Territorial governor Lewis Cass interceded in the matter, convincing Tiffin to revisit the matter.

Fortunately, a silver lining emerged as the negative opinions about Michigan Territory failed to gain widespread traction. In the wake of the federal government's decision to offer eighty-acre parcels of land for just $1.25 an acre in 1818, a steady influx of determined settlers began to make their way into the region. However, the path to settling in this promising territory was fraught with challenges, and overland transportation emerged as a significant impediment to western expansion.

Prospective settlers often had to trace the footsteps of Native Americans who had traversed the land before them. These trails were far from

Introduction

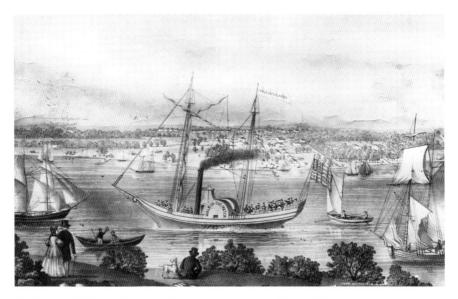

Depiction of *Walk-on-Water*'s maiden voyage to Detroit. *Library of Congress.*

accommodating. They meandered through rugged terrain, laden with deep ruts, formidable rocks and occasional stagnant pools of brackish water, making the arduous journey even more grueling.

The Great Lakes region would witness a transportation revolution on the morning of August 25, 1818. *Walk-on-Water*, the very first steamboat to sail the Great Lakes, departed Buffalo, New York, on its maiden voyage across Lake Erie. Twenty-nine passengers were aboard the paddle wheel–driven boat, although it could have accommodated several dozen more. With stops in Cleveland, Sandusky and Detroit, the round trip took nine days from and back to Buffalo, located at the eastern end of Lake Erie, at the head of the Niagara River, on the U.S. border with Canada.

The introduction of steamboats on Lake Erie had a profound impact on westward expansion in the United States. These steam-powered vessels revolutionized transportation, replacing unreliable wind-powered sailing vessels with erratic schedules. Lake Erie, a crucial gateway to the American West, saw increased access to rapidly growing western territories like Michigan, Indiana, Ohio and Illinois, stimulating westward migration. Steamboats became economic linchpins, facilitating the movement of people and goods, thus streamlining the transport of the Midwest's agricultural products and raw materials to eastern markets. This not only spurred economic growth in the western territories but also played a vital

Introduction

role in the region's prosperity. As steamboat traffic surged, port cities such as Detroit flourished, becoming pivotal hubs for commerce and creating opportunities for settlers, merchants and entrepreneurs, further fueling western expansion.

The Erie Canal

Even before *Walk-on-Water* and other steamboats traversed Lake Erie, another transportation advance was under construction. The Erie Canal, a monumental engineering feat begun in 1817 and completed in 1825, would play a pivotal role in shaping the development of the United States during the early nineteenth century. It was constructed to connect the Great Lakes region to the Eastern Seaboard, transforming transportation, trade and settlement patterns in the country.

The Erie Canal had a profound impact on Michigan Territory, particularly in facilitating the influx of settlers and the eventual attainment of statehood. This monumental waterway provided a crucial transportation link between the Midwest and the eastern United States. Michigan Territory, characterized

Erie Canal overview. *Library of Congress.*

Introduction

An early steamboat traversing the Erie Canal. *Library of Congress*.

by its vast natural resources and fertile farmland, was a promising destination for pioneers in the early nineteenth century. However, prior to the Erie Canal, reaching Michigan was a challenging and expensive endeavor due to the lack of efficient transportation routes. The canal changed this dynamic significantly.

Settlers from the eastern states found it easier and more cost-effective to travel to Michigan via the Erie Canal. They could embark on a journey by boat, passing through the canal and its connecting waterways to reach the Great Lakes, including Lake Erie. This reduced the time and effort required for overland travel, making Detroit and Michigan increasingly attractive destinations for those seeking new opportunities.

Moreover, the canal's impact extended beyond transportation. It played a pivotal role in boosting Michigan's economy. The improved access to eastern markets allowed Michigan's farmers and businesses to export their products more easily. Agricultural expansion, timber harvesting and mineral extraction thrived, contributing to economic growth in the territory.

The population growth and economic development catalyzed by the Erie Canal also had political consequences. Michigan's increased population and economic potential fueled a desire for statehood. In 1837, Michigan was admitted to the Union as the twenty-sixth state, in part due to the growth

An abandoned section of the Erie Canal. *Library of Congress*.

and prosperity facilitated by the canal. Thus, the Erie Canal not only eased the path for settlers to reach Michigan but also played a crucial role in the territory's economic development and eventual transition to statehood, shaping its history and future in profound ways.

A Revolution in Agriculture in Michigan

Throughout the nineteenth century, family farming was a pivotal part of Michigan's economy. In 1860, for example, 85 percent of the state's population depended on farming for its livelihood; one hundred years later, the Michigan countryside had so depopulated that fewer than one in four either lived in rural areas or depended on farming for its livelihood. Late nineteenth-century industrialization would bring about large, mechanized, corporate farming, as well as crop specialization that would have a devastating impact on traditional family farming.

Before the nineteenth century, the Indigenous peoples of Michigan practiced a mix of hunting, gathering and basic farming just to survive. Their common crops included corn, beans, peas, squash and pumpkins, but their agricultural footprint was limited, resulting in minimal impact or alterations to the landscape. In the seventeenth century, French explorers ventured into Michigan and discovered this pristine landscape. Subsequent French farming, albeit on a small scale, concentrated on cultivating fruit trees, particularly pear and apple varieties developed around Detroit. Agriculture, however, played a secondary role to the more profitable lumber and fur trades. As the fur trade eventually waned, and government policies shifted, agriculture gained prominence. Setting the stage for the growth and development of the family farm were the introduction of steamboats on Lake Erie and the completion of the Erie Canal in 1825, despite obstacles such as hostile fur traders, the threat of British rule and unfavorable land survey reports, not to mention the relocation of Native Americans west of the Mississippi by 1833.

The southern third of the state, blessed with fertile soil and accessible transportation routes, witnessed the earliest settlements. New Englanders were the predominant pioneers, finding the region ideal for wheat and wool production. Population growth was astonishing, with a tenfold increase between 1820 and 1834.

The Upper Peninsula's agricultural potential remained untapped until the mid-1800s, when it was introduced to support the burgeoning lumbering and mining industries. European immigrants arrived later than New Englanders, adding to the state's cultural diversity and introducing new agricultural practices.

Detroit's Strategic Advantage

Detroit's location in the nineteenth century was ideal not only as a point of access to the region's agricultural riches but also for business and commerce due to a combination of natural and geographical advantages. These factors propelled Detroit into a thriving economic center during this era.

Introduction

Geographic Location

Situated in southeastern Michigan, Detroit's strategic location along the Detroit River was a major asset. The river served as a crucial waterway connecting the Great Lakes region to the St. Lawrence River and, in turn, the Atlantic Ocean. This made Detroit a vital gateway to the western frontier and facilitated the transportation of goods. The city's proximity to the Great Lakes also allowed easy access to the Canadian market and beyond.

Abundant Natural Resources

The surrounding region was rich in natural resources. Vast forests provided an abundant supply of timber, essential for constructing buildings, ships and early factories. Additionally, the nearby mineral deposits, including iron and copper, played a pivotal role in the growth of industry. The fertile land in the surrounding areas supported agriculture, enabling the city to become a hub for food production.

Transportation Infrastructure

In the nineteenth century, Detroit invested heavily in transportation infrastructure. The city developed a network of roads and mile markers that allowed for the efficient movement of goods and people. The construction of railroad lines and spurs, like Milwaukee Junction, which at the turn of the century would be called the "cradle of the Detroit auto industry," further enhanced connectivity, facilitating the flow of raw materials to factories and finished products to markets.

Proximity to Trade Partners

Detroit's location on the U.S.-Canada border facilitated trade with British North America (now Canada). This cross-border trade was essential for the city's economic growth. The international boundary was porous, and goods flowed freely between the two regions, promoting commerce and cultural exchange. The city also served as a crucial trade center during the War of 1812, as it was a link between the western frontier and British-controlled Canada.

Introduction

Agricultural Hinterland

Detroit's location near fertile agricultural lands in the Midwest was instrumental in its role as an agricultural hub. Farmers from the surrounding region brought their produce to Detroit's markets, making it a center for agricultural trade and distribution. The city's agricultural hinterland extended into the fertile regions of Michigan, ensuring a consistent supply of foodstuffs.

Diversity among New Settlers to Detroit

In the realm of immigration theories, a prevailing concept suggests that the attraction of opportunities in a new place often outweighs the push of adversity back home, often summed up as "the pull being stronger than the push." This principle was clearly observed in the migration patterns to Detroit, where economic prosperity played a significant role in drawing people, rather than them being compelled by challenging circumstances in their home states or countries. This was especially true for New Englanders and New Yorkers, often referred to as "Yanks" and "Yorkers," who saw the chance to acquire fertile lands at affordable rates as a strong incentive. The opening of the Erie Canal facilitated this migration to southeast Michigan.

By the mid-nineteenth century, on the other hand, a large number of Irish immigrants felt compelled to leave their homeland due to the devastating potato famine. Similarly, many Germans left their country following a failed democratic revolution in 1848. Still, the largest single group of immigrants to the Detroit area consisted of Canadians with Irish, Scottish and British backgrounds. Immigration patterns would evolve in the last quarter of the nineteenth century, with the arrival of other immigrant groups from western Europe, including Italy and Greece, as well as immigrants from eastern Europe, particularly Poland, Hungary and Russia.

In Detroit during this era, individuals could achieve success in accordance with their skills and abilities and could climb the socioeconomic ladder due to the city's high demand for labor. Detroit, at that time, held the position as a mid-sized urban center that fostered a familial atmosphere, where homeownership was financially feasible. This welcoming environment was particularly beneficial for Catholic immigrants, like the Irish, given the city's existing Catholic presence with French roots.

Southeastern Michigan once stood as the third-largest settlement area for European immigrants. Immigrants to the Detroit area often grappled with the challenge of balancing assimilation into American culture while preserving their original heritage. First-generation immigrants were cautious about their children becoming overly Americanized, fearing a disconnection from their cultural roots and family ties.

Beginning as early as 1850, Black individuals began arriving in Detroit, spurred by the enactment of the Fugitive Slave Law, which forced people of color to flee the Deep South in pursuit of freedom in northern cities like Detroit. Before the turn of the century, Detroit also began witnessing an influx of individuals from the Middle East and Asia.

The motivations behind migration shifted over time. Before the Civil War, individuals sought Detroit for land opportunities. However, after the Civil War, the city's industrial economy experienced rapid growth, prompting many to relocate in search of employment opportunities. Some may have initially come for land but soon realized the appeal of factory jobs that offered, if not high wages, at least reliable and consistent compensation.

Detroit's First Industrial Era: Gone *and* Forgotten

The era spanning from 1850 to 1900 during Detroit's initial industrialization often fades into obscurity when juxtaposed against the city's monumental dedication and resources poured into establishing itself as a global hub for automobile manufacturing. Yet the multitude of smaller industries that emerged during this earlier period, only to later be overshadowed by the automotive giants, has regrettably been neglected by historians. Their pivotal role in fostering the rapid development of Detroit remains largely unexplored. It's crucial to recognize that the foundational elements essential for launching a complex industry like automobile manufacturing didn't suddenly materialize with Ransom Olds and Henry Ford.

The established presence of vital components such as iron and steel foundries, tool and die shops, carriage builders, gasoline engine manufacturers, skilled mechanics, wheel manufacturers and paint and varnish companies, among various other local industries, significantly fueled the subsequent explosion of the automotive sector and merits substantial acknowledgement. Even broader industries with seemingly distant connections to automobiles warrant historical investigation. They not only contributed to diversifying the economy but also provided stability, fostering a community of income-earning residents who formed a substantial market for diverse manufacturers, including the automobile industry. Consequently, exploring these diverse industries provides a more comprehensive understanding of Detroit's industrial evolution, showcasing the interconnectedness that underpinned its economic growth.

The stove manufacturing landscape in Detroit stands as a prime example. Jeremiah and James Dwyer recognized early on that they weren't reinventing the wheel in the cast-iron stove industry. The market was firmly established on the East Coast, yet midwesterners faced challenges ordering from afar and enduring endless waits for their desired product. Identifying this gap, the Dwyer brothers capitalized on it. However, what set them apart was their meticulous attention to detail, which garnered recognition from a clientele valuing the superior quality of Detroit-made stoves. By the conclusion of this initial phase of industrialization, the Dwyer enterprises had remarkably penetrated eastern and international markets.

Understanding how and why the stove industry flourished requires examining what drove its success, similar to any complete exploration of the automobile industry. This involves recognizing the key factors that fueled its growth. The significant discoveries of minerals in Michigan's Upper Peninsula played a pivotal role in igniting this industry. Concurrently, the construction of the Soo Locks marked a significant milestone, easing the transportation of vital materials to Detroit via waterways. Delving into these two pivotal catalysts that initiated Detroit's initial industrial era sheds light on why the stove industry, as well as many other contemporaneous Detroit industries, thrived in the city.

Chapter 1

TWIN DRIVERS OF DETROIT'S FIRST INDUSTRIAL ERA

Before Detroit could ever assert itself as an industrial powerhouse, it would need access to raw materials not found on the farm or in the bountiful forests throughout the state. To produce the products Detroit would become identified with—stoves, railroad cars, factory machinery, cast-iron products and eventually automobiles—the city would need access to large quantities of iron ore. During the mid- to late nineteenth century, mineral discoveries in the Upper Peninsula and dramatic improvements in transporting raw materials from Lake Superior to Lake Huron via the St. Mary's River would bring about a monumental transformation in the industrial capacity and product development within the city of Detroit.

The Discovery of Copper and Iron in the Upper Peninsula

Nestled in the heart of the Great Lakes basin, Michigan boasts an extraordinary wealth of natural resources. Its riches span fertile farmlands, vast timberlands and a diverse array of minerals formed over countless millennia through complex geological processes.

The Upper Peninsula in particular stands out for its rich mineral deposits, located in proximity to the shores of Lake Superior. This region is renowned for its historic reservoirs of iron ore and native copper. Over the past 175 years, these rugged landscapes have yielded prodigious quantities of these valuable minerals, beckoning early settlers to find employment in copper and iron mines and, in doing so, shaping the region's history.

Native Americans were the earliest pioneers in mining, ingeniously harnessing float copper remnants, which were left behind by retreating glaciers. They also employed rudimentary methods involving pit digging and the use of heavy stones to extract copper from surrounding rock. In both cases, they crafted various items such as bracelets, beads, tools and fishhooks for trade. Their craftsmanship extended the reach of these copper goods as far as the Mississippi River, as evidenced by their discovery at archaeological sites throughout the Mississippi watershed.

In 1668, Jesuit missionary Claude Dablon, dedicated to converting Native Americans in the upper Great Lakes region, stumbled upon the Ontonagon Boulder—a colossal 1.5-ton chunk of native copper—nestled along the Ontonagon River. Recognizing its significance, Dablon meticulously documented his discovery of copper-rich areas like this and relayed the information to his superiors in Canada.

Fast-forward to the early 1830s, when Douglas Houghton, a budding geologist and physician, spearheaded an expedition that delved into the copper deposits nestled within the terrain of the Keweenaw Peninsula. As Michigan gained statehood in 1837, Houghton, alongside other trained surveyors, conducted a comprehensive geological survey of the Upper Peninsula. His groundbreaking work earned him the distinction of becoming Michigan's inaugural state geologist in 1839. In his illuminating reports of 1841, Houghton spotlighted the vast potential of the Upper Peninsula, particularly emphasizing its abundant copper resources.

By the early 1840s, a series of events had ignited significant interest in exploring the mining potential of the Keweenaw Peninsula. American prospectors stumbled upon numerous chunks of native copper strewn across the ground or in nearby streams. Additionally, the Treaty of LaPointe in 1843 between the United States and the Ojibwe (Chippewa) Indigenous peoples marked a pivotal moment. This treaty involved the cession of extensive lands in Wisconsin and Michigan, including the mineral-rich western Upper Peninsula (the eastern portion of the U.P. had already been ceded to the United States in 1836 following the Treaty of Washington). In exchange, the Ojibwe retained rights to hunt, fish and gather on these ceded territories, with the United States obliged to provide reservation land for Native American signatories.

In the spring of 1843, Captain Walter Cunningham was appointed as a special agent to the Keweenaw Peninsula by the United States. He established the region's first mineral land agency, known as the Government House, near present-day downtown Copper Harbor.

These significant events paved the way for the discovery of substantial copper deposits in the Keweenaw Peninsula by the mid-1840s. This marked the beginning of "copper fever," a fervent rush that attracted numerous American and immigrant prospectors eager to stake claims and pursue economic opportunities on the burgeoning mining frontier.

Another surveyor of note was William Burt, who not only played a crucial role in the exploration and identification of mineral resources in the Upper Peninsula but also, as an inventor, made significant contributions to the field of surveying. Burt had moved to Michigan from Massachusetts in the 1830s to pursue a career in surveying and sawmill construction around the state. Perhaps Burt's most notable achievement, outside of documenting the vast iron ore deposits in and around Negaunee, was the invention of the solar compass in 1835. This innovative instrument used the sun's position to determine direction, eliminating the need for a magnetic compass. The solar compass proved particularly useful in areas where magnetic variations could affect traditional compasses, such as regions with substantial iron deposits, like Michigan's Upper Peninsula. His solar compass proved instrumental in accurately surveying the Upper Peninsula and elsewhere.

In the autumn of 1844, during the process of subdividing multiple U.P. townships, Burt and his sons encountered a barrage of challenges: inclement weather, scarcity of supplies and the rugged, unforgiving terrain characteristic of Michigan's Lake Superior coast. Despite these obstacles, Burt remained resolute, urging his sons to maintain their focus and eschew distractions. He was unwavering in his commitment to ensure the precision of his reports, refusing to sacrifice accuracy for expediency. The integrity of his field notes was paramount. As they continued their measurements within the area known to the Ojibwe as Negaunee, meaning "foremost" or "pioneering," the conventional magnetic compasses wielded by his sons were

William Burt (1792–1858). *Chares Tuttle*, General History of the State of Michigan.

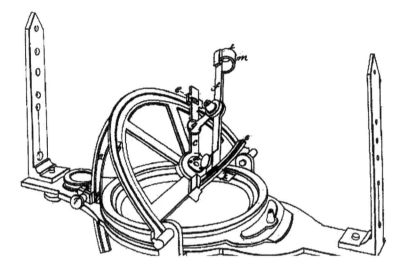

Patent diagram of William Burt's solar compass, 1836. *Charles Tuttle,* General History of the State of Michigan.

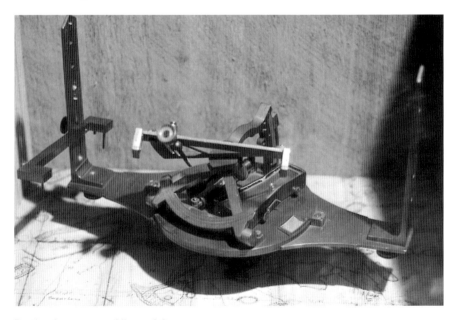

Burt's solar compass. *Library of Congress.*

thrown into disarray. The needles floated erratically, swaying left, then right, disrupting any attempts at establishing precise line placements. In response, Burt instructed his sons to fan out and search for ground irregularities that could account for this magnetic disturbance. It didn't take long to deduce that the cause lay in the abundance of iron ore outcroppings within close proximity. With the adoption of his solar compass, Burt's team painstakingly completed the survey, achieving a level of accuracy far surpassing their previous efforts. Perhaps most significantly, it was during this endeavor that William Burt and his sons made the formal and official discovery of iron ore in Michigan.

This discovery led to the establishment of the Marquette Iron Range, one of Michigan's major iron regions. Prospectors would continue to discover additional iron ore deposits in Michigan, including the Eastern Menominee Range in 1845, the Western Menominee Range in 1851 and the Gwinn District of the Marquette Range in 1869. In stark contrast, the earliest iron ore discoveries in Minnesota did not materialize until over three decades later, underscoring the pioneering role that Michigan played in the nascent iron ore industry. Michigan firmly established itself as a cornerstone contributor to the nation's mineral wealth.

It was the allure of iron and copper that ushered in the first major population boom in the region. The initial wave of immigrants to the Upper Peninsula comprised the Cornish, who brought with them centuries of mining expertise, followed by Germans, Irish and French Canadians. In the latter part of the century, immigrants from Italy, Finland, Scandinavia, Poland, Russia, the Austro-Hungarian Empire, Wales, Scotland and even as far as the Isle of Man and China made their way to the area. These diverse groups infused the region with their ethnic traditions and culinary heritage. By 1917, traveling from Houghton to Calumet, a mere twelve miles apart in the Keweenaw Peninsula, felt like entering a foreign land. Ethnic churches, newspapers, clubs and shops thrived in communities where over 75 percent of the population hailed from foreign shores. This immigrant tradition left an indelible mark on the region, fostering a rich tradition of dialectic folklore, as noted by folklorist Richard Dorson.

Initially, the discovery of iron ore along the northern coast of Michigan's Upper Peninsula could be seen as a boon more to the Canadian economy than to Michigan's lower peninsula, much less Detroit. This wealth of resources simply wasn't accessible to areas like Detroit and the lower Great Lakes. Their direct access to such raw materials remained obstructed, primarily because the passage through the St. Mary's River—connecting

Lake Superior to Lake Huron—proved to be a treacherous route for ships of varying sizes. This hurdle posed a significant barrier, preventing Detroit from capitalizing on the abundant mineral ore findings until a viable solution could be devised to navigate this challenging waterway.

The Opening of the Soo Locks

During the late nineteenth century, a crucial component of the Great Lakes transportation system developed out of commercial and industrial necessity. The construction and operation of the Soo Locks, located in Sault Ste. Marie, Michigan, have played a pivotal role in shaping the economic and industrial landscape of Michigan's Upper Peninsula.

The origins of the Soo Locks can be traced back to the early nineteenth century, when the United States and Canada began to recognize the immense economic potential of the Great Lakes as a transportation corridor. During the 1840s, significant copper deposits were discovered in the Keweenaw Peninsula, and shortly thereafter, substantial iron ore deposits were extracted in the vicinity of Negaunee, situated near the shores of Lake Superior, the largest among the Great Lakes. These findings represented pivotal moments, especially since iron ore, in particular, became an essential raw material for Detroit's rapidly growing industrial economy. However, there was a significant obstacle to realizing this potential—the St. Mary's River Rapids.

The St. Mary's River, connecting Lake Superior to Lake Huron, presented a formidable challenge with its treacherous rapids and cascades. In the river's upper region there is a notable drop of twenty-one feet over rapids below, which was a compact sandstone terrain stretching nearly three-quarters of a mile. The abrupt change in water elevation, referred to as the "Sault" in homage to its original French terminology, presented a formidable obstacle for trade vessels. These rapids rendered it impossible for such vessels to navigate through smoothly. As a result, a laborious process ensued, where cargo-laden ships were compelled to undergo unloading, followed by a challenging overland portage of their contents, only to be reloaded again.

Early attempts to address the navigation challenge of the St. Mary's River Rapids were modest in scale and lacked the resources and technology for a comprehensive solution. Various canals and locks were proposed, but they were often beset by financial difficulties and technical limitations. The year 1797 marked a turning point when the first lock of the St. Mary's River

LOOKING DOWN—ILLUSTRATES THE HEIGHT OF THE LOCKS.

Soo Locks. *Library of Congress*.

emerged along the northern shoreline. This lock was specifically designed to facilitate the passage of trade canoes, marking a pioneering effort to overcome the natural impediments posed by the river's formidable rapids. However, this early engineering achievement faced a grim fate during the War of 1812, as American forces obliterated it. In the aftermath of this destruction, the onerous process of unloading, hauling cargo overland and reloading vessels became the norm once more.

As demand for iron ore and other goods from the U.P. surged in the late nineteenth century, the urgency of overcoming this challenge became increasingly apparent. There was a growing demand for a navigational channel that would allow vessels to safely transport iron ore and other commodities from Lake Superior to Lake Huron.

The Soo Locks project conducted by the State of Michigan emerged as a complex and ambitious solution to this challenge. In 1855, the first lock, known as the State Lock, was completed. However, it was apparent that a larger and more efficient lock was needed to accommodate the growing size of vessels and the increasing traffic in the Great Lakes.

The most significant milestone in the development of the Soo Locks came about with increased federal involvement and funding. The U.S. Army Corps of Engineers took charge of the project, recognizing its national significance. For the mineral wealth of the U.P. to reach Detroit and various eastern ports, a new lock had to be designed to be large enough to accommodate the largest vessels on the Great Lakes. The Poe Lock, named after Orlando Poe, the chief engineer of the Corps of Engineers, was considered revolutionary at the time and ensured that massive vessels could safely navigate the St. Mary's River. Construction of the Poe Locks took fifteen years, beginning in 1881 until its momentous opening in 1896. The dimensions of the Poe Locks were 1,350 feet in length and 80 feet in width. The newest lock marked a new era in Great Lakes transportation and catalyzed the region's economic transformation, interconnecting the vast natural resources of the U.P. with Detroit to the south and other industrial centers to the east.

Once such a safe and efficient method of transporting iron ore and other raw materials became possible, mining operations throughout the Marquette and Gogebic Ranges expanded, creating jobs, attracting a skilled workforce and stimulating the local economy. The city of Sault St. Marie, where the locks are located, saw significant growth as it became a hub for shipping and trade.

The Soo Locks played a pivotal role in Detroit's industrial expansion. The newfound ease of transporting critical raw materials facilitated by the locks allowed the city's manufacturing sector to flourish. Iron ore, coal and timber could now be efficiently shipped from the upper Great Lakes region to Detroit. Accessibility underpinned the growth of Detroit's industrial base, including the manufacture of stoves, railroad cars, large and small boats, cast-iron products, factory machinery and steel.

Furthermore, the Soo Locks contributed to the expansion of Detroit's maritime industry. The city became a major hub for shipbuilding and repair, creating jobs and fostering a vibrant maritime community. This growth in the maritime sector complemented Detroit's growing manufacturing capabilities, contributing to the city's economic prosperity.

In the years to come, the Soo Locks would prove to be of vital national importance. During World War II, for example, the locks played a crucial role in the movement of iron ore and other strategic materials. The ability to move these materials efficiently was essential to the war effort, as steel production was critical for the manufacture of weapons and equipment. Recognizing their strategic value, the U.S. government invested in the modernization and expansion of the Soo Locks to ensure their continued operation.

Chapter 2

VISIONARIES AND ENTREPRENEURS WHO BUILT DETROIT IN THE LATE NINETEENTH CENTURY

In 1840, Detroit stood on the cusp of a remarkable transformation, shifting from a rustic frontier town with a modest population of 9,102 to a bustling nexus of transportation and industry. The year 1850 witnessed an astonishing surge, with the population more than doubling to 21,019, and by 1860, this figure had doubled once again, reaching 45,619. This meteoric rise was fueled by two pivotal catalysts—the discovery of iron ore and the opening of the Soo Locks—that propelled Detroit into its role as an emerging industrial juggernaut.

In 1860, Detroit featured several blast iron furnaces, two railroad car manufacturing plants, seven copper smelters, a brass foundry, a dry dock dedicated to shipbuilding and several tobacco product manufacturing facilities. Copper smelting was the most valuable manufacturing industry in the Detroit–Wayne County area, with sawed lumber a distant second. The city also featured an array of businesses, including a few dozen breweries, fifteen tanneries and six flour mills. Meatpacking would also become a major industry in Detroit during this period. Construction overall saw a significant surge during this time. By 1869, Detroit proudly claimed a total of 16,152 buildings within its borders. Remarkably, almost one-fifth of this impressive count had sprung up in just the preceding two years.

While copper and timber production were the primary economic pillars in 1860, Detroit industry underwent a transformative shift. As the mid-decade mark approached, the manufacturing landscape saw a remarkable evolution, with steel, iron and foundry products taking center stage. During this pivotal period, several burgeoning enterprises emerged as leaders in this industrial renaissance. Detroit Bridge and Iron Works, the Detroit Safe Company, the Michigan Car company and the E.T. Barnum Wire and Iron Works were all newly established ventures that would rapidly ascend to prominence during the latter half of the decade.

The most significant industrial establishment at this time was the Eureka Iron and Steel Works, strategically situated nine miles downstream from Detroit and under the ownership of Eber Brock Ward, Detroit's pioneering millionaire industrialist. As the Civil War unfolded, this establishment emerged as an important regional employer, boasting a workforce of over 250 individuals. Strategically perched along the banks of the Detroit River, the Eureka Iron and Steel Works enjoyed an advantageous position to efficiently receive substantial shipments of newly discovered iron ore from the U.P. via maritime routes. Remarkably, many of these essential raw materials were transported in vessels also owned by Ward himself. In a groundbreaking development in 1864, Ward's facility, which happened to be the largest in Wayne County, achieved a historic milestone by becoming first in the United States to commercially produce steel using the revolutionary Bessemer method. This innovation marked a significant leap forward in terms of cost-effective mass production of steel, as it transformed pig iron into high-quality steel on an unprecedented scale.

Detroit's highly successful stove industry, which would directly benefit from the mineral discoveries and their availability, still required vision and leadership to overcome the market penetration of eastern stove brands. Jeremiah and James Dwyer were two such visionaries and entrepreneurs up to the challenge.

Detroit was privileged to witness the rise of a cadre of such forward-thinking visionaries and trailblazing entrepreneurs, pivotal in constructing the city's industrial infrastructure before the automobile era. While the nineteenth century welcomed tens of thousands of newcomers to the city, only a select few distinguished themselves as dynamic leaders in both industry and public service. Regrettably, their contributions have often been overlooked in the annals of pre–automobile industry history, warranting acknowledgement, albeit brief, in this narrative.

Eber Brock Ward:
Detroit's First Millionaire Industrialist

Eber Brock Ward, often referred to as E.B. Ward, was a prominent figure in the industrial landscape of Detroit. Born on December 25, 1811, in Waterloo, Ontario, to American parents, Ward relocated with his family to Detroit in 1821. At a young age, he embarked on a maritime journey, working as a cabin boy and deckhand, which exposed him to the intricacies of the shipping industry.

Initially, Ward invested in a vessel named the *General Harrison*, in which he held a 25 percent ownership stake. Eventually, he became the master of this vessel in 1835, demonstrating his proficiency in its operation. His success led him to form a partnership with his uncle Samuel Ward, a notable figure in the maritime industry at Marine City, Michigan.

Under this partnership, Ward continued to thrive, participating in various shipbuilding projects, both steamers and sailing ships. Some of the notable vessels included the *Samuel Ward, Detroit, Champion, Pacific, Caspian, Ocean, Huron* and others.

In 1850, Ward decided to relocate his operations to Detroit, where he became deeply involved in the shipbuilding business. His enterprises played a pivotal role in the construction of numerous vessels, further solidifying Detroit's significance as a hub for shipbuilding and maritime activities.

Ward's business acumen extended beyond shipbuilding, as he recognized the importance of both the discovery of iron ore in the Upper Peninsula and the engineering marvel that was the Soo Locks. In 1853, he established the Eureka Iron Works, situated in the downriver community of Wyandotte. The facility became the first in the United States to successfully implement the Bessemer process for steel production, revolutionizing the steel industry by enabling efficient mass production.

Simultaneously, Ward acquired extensive timberlands along the Pere Marquette River in Lake County, near Ludington, starting around 1852. He patiently waited for the timber to mature, strategically positioning himself in the burgeoning logging industry.

Ward's influence expanded further when he assumed the presidency of the Flint and Pere Marquette Railroad Company in 1860. This role allowed him to control the transportation of his timber, ensuring the efficient movement of raw materials to his various enterprises.

In 1869, Ward made a significant land purchase, securing a vast tract of seventy thousand acres in Ludington's fourth ward, alongside Lake

Pere Marquette. Here, he erected a state-of-the-art sawmill known as the North Mill perched on fifty-five stone piers and equipped with cutting-edge technology. This impressive mill had a daily capacity of 100,000 board feet.

Ward's commitment to his industrial pursuits was unwavering, and he continued to expand his operations. In 1871, he constructed a warehouse near the original mill, facilitating the storage and sale of supplies to his employees. Subsequently, Ward built another mill in proximity, which earned the distinction of being considered the finest sawmill in the United States.

Throughout his lifetime, E.B. Ward was not only an astute businessman but also a philanthropist, contributing significantly to various cultural and social institutions. His legacy endured, and he played a pivotal role in shaping Detroit into a major industrial center during the nineteenth century. Ward's dedication to diversifying his investments and fostering innovation made him Detroit's first millionaire, leaving an indelible mark on the city's industrial history. A recent book by Michael W. Nagy, *The Forgotten Iron King of the Great Lakes* (2022), sheds a very bright light on Ward and his considerable influence.

Dr. Samuel Duffield and the Origins of Parke-Davis

Parke-Davis, one of the pioneering pharmaceutical companies in the United States, traces its origins to a small drugstore on Woodward Avenue in Detroit. Founded by Dr. Samuel P. Duffield in 1866, the store operated under the name of S.P. Duffield & Company. In its early years, the company focused primarily on producing simple pharmaceutical products, such as elixirs, tinctures and herbal remedies. As the business expanded, Duffield found it necessary to partner with Hervey Coke Parke, who brought considerable organizational skills to the enterprise. In 1867, George S. Davis purchased an interest in the firm of Duffield, Parke & Company, manufacturing pharmacists. Two years later, Duffield left the firm to resume the practice of medicine. By 1875, the firm had formally incorporated as Parke, Davis & Company.

Parke-Davis made a significant breakthrough in the late nineteenth century when it introduced the concept of standardization in pharmaceuticals. This was a pivotal moment in the company's history and the pharmaceutical industry as a whole. Parke-Davis pioneered the practice of accurately

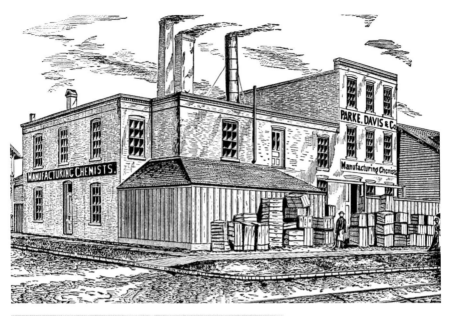

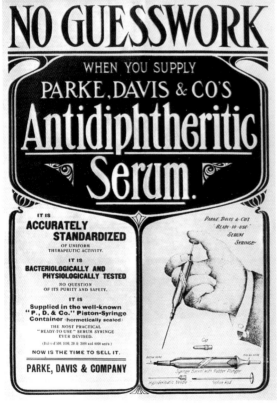

Above: Parke-Davis manufacturing facility on the Detroit riverfront. *Author's collection.*

Left: An advertisement for anti-diphtheria medication, Parke-Davis Pharmaceuticals. *Author's collection.*

measuring the active ingredients in its medications, ensuring consistency and effectiveness. This innovation set a new standard for quality and safety in pharmaceuticals, and it became a hallmark of the company's reputation.

One of Parke-Davis's most famous contributions to the pharmaceutical industry was the development of a stable and standardized form of digitalis, a plant extract used to treat heart conditions. Digitalis had been used for centuries, but its efficacy was inconsistent due to variations in the potency of different plant specimens. Parke-Davis's standardized digitalis revolutionized the treatment of heart diseases and became one of its flagship products.

Another groundbreaking achievement came in 1889, when Parke-Davis became the first American company to produce and market insulin, a lifesaving treatment for diabetes. This innovation, developed in collaboration with Canadian scientists Frederick Banting and Charles Best, was a testament to the company's commitment to advancing medical science and improving patient outcomes.

Parke-Davis continued to expand its portfolio of pharmaceutical products, researching and manufacturing medications to address a wide range of medical conditions. The company's dedication to research and development led to breakthroughs in areas such as anesthesia, vaccines and antibiotics.

By the early twentieth century, Parke-Davis had established itself as a global leader in the pharmaceutical industry. It played a pivotal role in shaping modern pharmacology and contributed significantly to the development of safe and effective medicines. Its commitment to rigorous scientific research, product standardization and collaboration with leading scientists positioned it as a trusted name in healthcare.

In 1970, Parke-Davis merged with Warner-Lambert, creating Warner Lambert/Parke-Davis, which later became part of Pfizer in 2000. The legacy of Parke-Davis lives on through its contributions to pharmaceutical science, its pioneering spirit in drug development and its dedication to improving the well-being of patients worldwide.

Hazen S. Pingree: A Visionary Leader and Rooter of Corruption

Hazen S. Pingree, a name etched into the annals of Detroit's history, was a visionary leader whose impact on the city and its people resonates to this day. Born on August 30, 1840, in Denmark, Maine, Pingree's journey from a humble upbringing to the mayorship of one of America's largest cities is a testament to his determination, innovation and unwavering commitment to the welfare of the common man.

Pingree's early life was marked by hardship and perseverance. Raised in a farming family, he learned the virtues of hard work and thriftiness from an early age. These values would later become the cornerstones of his political career. In 1863, he moved to Detroit, where he started as an apprentice shoemaker and eventually built a successful shoe manufacturing business, demonstrating his entrepreneurial spirit and astute business acumen.

In 1866, Pingree and his partner Charles H. Smith acquired Henry Baldwin's shoe company, where Pingree had previously worked. Over the next two decades, they expanded the company into a million-dollar enterprise, employing over seven hundred workers and producing more than half a million shoes and boots annually. It became the Midwest's largest shoe manufacturer and the second largest in the entire United States.

Tragically, in 1887, a fire ravaged their manufacturing facility, causing complete destruction. However, they were able to gradually, though not completely, recover.

Today, Pingree Detroit stands as a testament to Pingree's legacy. Established in 2015 as a worker-owned cooperative, it specializes in crafting and selling handcrafted bags, home and pet goods, accessories and footwear. Utilizing high-quality leather reclaimed from Detroit's auto industry, Pingree Detroit continues Pingree's spirit of innovation and production excellence. Pingree would likely take pride and joy in the continuation of this tradition.

Pingree's entry into politics came in 1889, when he was elected mayor of Detroit. It was during his time in office that he became a symbol of progressive governance and social reform. His administration was characterized by innovative policies aimed at improving the lives of the city's residents. One of his most iconic initiatives was the "Potato Patch" campaign. Facing a recession and widespread unemployment, Pingree encouraged Detroiters to plant vacant lots with potatoes, creating a source of both food and employment. This innovative idea earned him the nickname "Potato Patch Pingree" and captured the hearts of the city's working-class population.

Pingree's commitment to social justice extended beyond the potato patch. He implemented a system of public ownership for utilities, which led to more affordable gas and electric services for Detroit's citizens. He also advocated for improved public transportation and initiated the construction of miles of new roads, laying the foundation for the city's modern infrastructure.

Detroit mayor Hazen Pingree at the plow, breaking ground for the construction of Grand Boulevard, August 10, 1891. *Burton Historical Collection, Detroit Public Library.*

Education was another cornerstone of Pingree's reform agenda. He worked tirelessly to improve Detroit's public schools, emphasizing the importance of accessible and quality education for all children, regardless of their economic background. His dedication to education earned him the title of the "Education Mayor."

In addition to his domestic reforms, Pingree was also known for his efforts to curb corruption in city government. He initiated a series of reforms to make city government more transparent and accountable, setting a precedent for ethical governance that would endure in Detroit's political landscape.

Beyond his mayoralty, Pingree's influence extended nationally. He was a staunch advocate for the rights of farmers and a vocal critic of corporate monopolies, earning him a reputation as a populist leader. His progressive ideas resonated with many, and he used his platform to champion reforms on a broader scale.

Hazen S. Pingree's life and legacy are a testament to the power of visionary leadership and the impact one individual can have on a city and its people. His innovative and compassionate approach to governance left an indelible mark on Detroit, transforming it into a more equitable and prosperous city. Pingree's legacy lives on in the hearts and minds of those who continue to be inspired by his commitment to the common good and his unwavering dedication to the people he served.

Mayor Christian Buhl
and Detroit's Central Farmers' Market

In April 1861, as Confederate troops were firing on Fort Sumter in Charleston Harbor, marking the commencement of the Civil War, Detroit underwent an unofficial transformation from a frontier town to a significant American city. The official inauguration of Detroit's new Central Farmers' Market by Mayor Christian Buhl not only fulfilled a campaign promise made nine months prior but also recognized Detroit as a vibrant urban center and not just a raw material processing hub. By 1860, Detroit's population had surged to 45,619, more than double its size from a decade earlier. Over the next three decades when the Central Farmers' Market was in operation, the city's population would grow to over 200,000 residents. For the first time, suburban farmers no longer needed to make the trek to the river docks to export their farm products for eastern markets; they could find a robust local market for their fruit, vegetables, eggs, meat and other agricultural goods all in one centralized location.

The Central Farmers' Market covered more than two square blocks, extending from Woodward Avenue to Randolph Street, adjacent to Campus Martius. At the heart of this public market stood the Vegetable Building, which was nearly the length of a football field. Local architect John Schaffer was commissioned to design an eye-catching open-air pavilion. The imposing slate roof was supported by prefabricated timber trusses resting on forty-eight cast-iron columns, meticulously molded to resemble stone. The cost to the city was $5,312, with an additional $1,500 for the slate roof. Cast iron had become a popular, durable material for both building and decorative purposes, a trend that would sustain various industries in the decades to come. Soon, an additional structure, known as the Market Building, primarily for the sale of meat and meat byproducts, would be added behind the Vegetable Building.

The Central Farmers' Market swiftly evolved into a community gathering hub. Beyond being a source of fresh food for local residents, the market area attracted entrepreneurs testing product ideas; factory managers and commercial establishments seeking workers; and friends, families and neighbors who came to congregate and socialize.

For over three decades, the Farmers' Market served as the epicenter of communal and commercial activities in the city of Detroit. In the early 1890s, several factors led the city administration to reconsider not the effectiveness of a public market but its current location. Increasing food spoilage and inadequate waste disposal in the area brought about a rat control problem. Additionally, the city council decided it was time to open up the area to accommodate the growing traffic. A new area to the east, subsequently named Eastern Market, eased concerns that various meats and fresh produce would vanish from the vicinity.

Eventually, the open-air pavilion was relocated to Belle Isle and repurposed for multiple uses, including housing police horses. More recently, through a collaboration with the Henry Ford, the structure was moved once again, this time to Dearborn. Following a decade-long funding initiative, the Vegetable Building has been meticulously restored and transformed into an educational venue so that future generations visiting Greenfield Village can appreciate this historical artifact of the region.

Captain Stephen Kirby and the Detroit Dry Dock Complex

The Detroit Dry Docks Engine Works complex, often overshadowed by Detroit's automotive legacy, holds a storied past. Perched on Atwater Street, this unassuming square building conceals a rich history, as it represents the last surviving fragment of a once-expansive shipbuilding complex that played a pivotal role in the development of Great Lakes maritime trade.

The tale begins in 1869, when the Dry Dock Engine Works was founded as a marine steam engine firm, operating out of a cluster of buildings at the intersection of Orleans and Atwater Streets. This marked the beginning of an era of expansion that extended through the late 1800s and early 1900s, eventually encompassing six interconnected buildings.

This complex, a hub of industry and innovation, played host to the construction and repair of numerous iconic vessels. It served as the birthplace or workplace for a distinguished fleet of ships, including the *Ste. Claire*, the *Columbian*, the *Greater Detroit* and the *Greater Buffalo*. Notably, it was also the site where a young machinist named Henry Ford honed his skills in the early 1880s, a glimpse into the region's transition from maritime prominence to automotive dominance.

Prior to the ascendancy of the Detroit Dry Dock, Marine City had been the state's primary shipbuilding hub. However, its remote location, distant from a skilled workforce, posed challenges. In 1872, Captain Stephen Kirby incorporated the Detroit Dry Dock Co., locating it at the corner of Atwater

Detroit Dry Dock marine engine repair facility. *Library of Congress.*

and Orleans Streets, a few blocks east of the present-day Renaissance Center. The *Detroit Free Press*, at the time, praised the company for offering "all the necessary materials for the repair of vessels and shipping, and no delay will occur in making vessels ready for service."

In August 1871, Frank Kirby, Stephen's son, relocated to Wyandotte, joining forces with his older brother Fritz to enter the world of shipbuilding. Their impressive skills caught the attention of Eber Brock Ward, a renowned shipbuilder and iron manufacturer at the time. Ward was so impressed that he proposed a partnership to fulfill a long-held vision: constructing a top-notch facility for building steel ships. His directive was simple: "Acquire the best materials and craft the finest ships."

Under the careful, capable and energetic leadership of the Kirby brothers, they assembled and organized the necessary equipment and materials. The first project after completing the shipyard was the construction of the massive 550-ton tugboat *E.B. Ward, Jr.* This vessel was acquired by the government and transported to the Eastern Seaboard, where it gained widespread admiration as a testament to its designers and builders.

In 1871, after E.B. Ward's passing, the Detroit Dry Dock Company purchased the shipbuilding facility. Frank Kirby assumed the role of superintendent, with his brother Fritz as the engineer. Over the subsequent years, the company not only built numerous vessels in Detroit but also produced a fleet of iron and steel steamships and propellers that rivaled the older and larger shipyards on the coast.

Until 1922, the Detroit Dry Dock Company operated two yards, producing a wide array of vessels, from lake steamers such as the *City of Detroit* to the 316-foot railroad ferry *Lansdowne*, constructed in 1884. A century later, the *Lansdowne* found a new purpose near Hart Plaza, serving as a floating restaurant. Frank Kirby also left his mark by designing the excursion steamers *Sainte Claire* and *Columbia*, which plied the waters for the Bob-Lo Island run until the late 1980s.

The Dry Dock Engine Works, situated on Atwater Street, contributed significantly to the shipbuilding industry. It manufactured 129 steam engines for marine use, and over one-third of these engines found their way to the Detroit Dry Dock Company. Interestingly, it was at the Dry Dock Engine Works that a young Henry Ford learned the machinist's trade during his tenure from 1880 to 1882. Julius Melchers, one of Detroit's early prominent artists, added the finishing touch to some vessels by crafting figureheads.

As the enterprise expanded, the construction of Dry Dock No. 2 in 1892 marked a milestone. This new facility became the second-largest dry dock on the Great Lakes, surpassed only by Port Huron's, capable of accommodating fully loaded ships for repairs, measuring 378 feet and supported by more than two thousand pilings.

With its increasing prominence, the operation managed to outmaneuver competitors, even confounding politicians. When Detroit mayor Hazen Pingree sought to establish an electric light plant on 350 feet of city-owned waterfront, he discovered that Dry Dock had already secured a twenty-year lease for the property, paying $1,300 per year. As a result, the plant had to be relocated elsewhere.

The Detroit Ship Building Company, a new entity formed by the consolidation of the Wyandotte yards and Orleans Street facilities, emerged as a successor to the Dry Dock Company in 1899. It would eventually evolve into AmShip Detroit. However, the prosperity that the company experienced during World War I waned, and the economic downturn of the 1920s took its toll.

The Wyandotte yard ceased operations in 1922, followed by the Orleans Street location seven years later. Today, the remnants of the Detroit complex, along with the remaining buildings, are commemorated in St. Aubin Park with a bronze monument depicting the 241-foot *Pioneer* in dry-dock comfort, evoking memories of Detroit's era as a vital shipbuilding center.

George Miller and the Beginning of Detroit's Thriving Tobacco Industry

The tobacco manufacturing history in Detroit spans an impressive 125-year period, commencing with George Miller's pioneering venture in 1841. His modest tobacco facility was housed in a small frame building, situated across from the historic Mariner's Church on Jefferson Avenue. Remarkably, Miller powered his operation using an aging, blind horse. The culmination of this era arrived in 1966, when the last tobacco facility, Schwartz-Wemmer-Gilbert, closed its doors.

Back to 1890, fifty years after George Miller began manufacturing cigars in his tiny facility on Jefferson Avenue: Detroit's many new tobacco manufacturers had collectively sold more than thirteen million pounds of chewing and smoking tobacco, along with just under eighty million cigars. These products primarily served the markets in the Midwest and on the East Coast. The following year, 1891, marked a significant milestone as tobacco manufacturing emerged as Detroit's leading industry.

By 1913, the tobacco manufacturing sector had solidified its prominence in the city's industrial landscape. It ranked first in the number of establishments, third in the number of people employed and fifth in terms of the value of its products. An intriguing aspect of this industry was its reliance on female labor, with the largest ten companies employing 302 men and 3,896 women, many of whom were under the age of twenty.

In 1925, a striking statistic highlighted the industry's vitality, with reports indicating that more than one million cigars were crafted in Detroit every day. Notably, all quality cigars were hand-rolled, which made this occupation one of the highest-paying jobs for women in the city, earning a weekly wage ranging from twenty-five to forty dollars. This significantly contributed to the economic empowerment of women.

Several tobacco companies emerged as prominent players in this thriving landscape, contributing to Detroit's reputation as a hub for tobacco manufacturing. Among the most successful firms were Banner, Bagley, Globe, Eagle, Daniel Scotten, Lilies, Mayflower and San Telmo. Each of these companies played a vital role in the city's industrial history. Banner, with its distinct quality and market presence, contributed to the growing reputation of Detroit's tobacco products. Similarly, Bagley's offerings were sought after, adding to the city's prominence in the tobacco manufacturing sector. Globe, Eagle and Mayflower also made significant strides in the industry. Their products not only catered to local markets but reached consumers far beyond Detroit's borders as well. These companies became synonymous with quality and craftsmanship in the world of tobacco. The success of Daniel Scotten, Lilies and San Telmo further underlined the resilience and innovation within Detroit's tobacco manufacturing. Their ability to adapt to changing market dynamics and consistently deliver exceptional products allowed them to maintain a strong foothold in the industry.

In conclusion, the history of tobacco manufacturing in Detroit is a testament to the city's industrial prowess and adaptability. From humble beginnings in the nineteenth century, the industry grew to become a cornerstone of Detroit's economy. It provided employment opportunities for women and created a reputation for high-quality products, which continue to be remembered as part of the city's rich heritage. The names of Banner, Bagley, Globe, Eagle, Daniel Scotten, Lilies, Mayflower and San Telmo—and more than two dozen other Detroit tobacco companies, large and small—echo through history, representing the legacy of Detroit's tobacco manufacturing industry.

Berry Brothers Paints and Varnishes

For over seventy-five years, the Berry Brothers brand stood as a global leader in paint and varnish manufacturing. Originating from modest roots, like many thriving industries in Detroit, Berry Brothers' story began when Joseph Berry, at seventeen, left New Jersey for Detroit and took on a clerk position at T.H. Eaton, a wholesale drug distributor. This role demanded long hours, with orders accepted as late as 9:00 p.m. and required for pickup by 6:00 a.m., compelling Berry to sleep on-site to fulfill these demands. Despite meager wages, this experience fueled his determination to forge his own success.

Gaining insights into chemicals and plant properties from his job, Berry developed a keen interest in varnishes made from natural substances. Armed with a small copper kettle, barely larger than a hat, he embarked on experiments with gums, resins and oils. After persistent trial and error, he crafted a remarkably durable surface varnish, finding a local market for his creation.

Relocating his enterprise to Springwells before its annexation by Detroit, Berry initially found commercial success with varnishes. However, his ongoing experimentation broadened his product range. When his older brother, Thomas, arrived from managing their father's tanning business

Trademark, Berry Brothers varnishes. *Author's collection.*

in Virginia, he joined Joseph, formalizing their partnership in 1858 and establishing the Berry Brothers brand for varnish products.

Outgrowing their Springwells facilities quickly, the brothers shifted operations to Leib and Wight Streets near Detroit's near east side riverfront. Originating from a small frame building with limited resources, the company embarked on an expansion campaign to meet demand and diversify their product line—spanning varnishes, shellacs, stains, paints and coatings for diverse applications like floors, moldings, carriages (both horse-drawn and motorized), pianos, organs and metal surfaces.

Expanding their reach, Berry Brothers established branches in major U.S. cities and, for European clients, in Antwerp. In 1903, recognizing the need, they inaugurated a separate production facility in Walkersville, Ontario, to cater to their Canadian market.

In 1910, three years after Joseph Berry's passing, Detroit hosted a convention for the International Association of Master Painters and Decorators of the United States and Canada. During a factory tour, the plant manager unveiled a significant relic: Joseph Berry's original copper kettle, the humble genesis of their business. Despite achieving immense personal success, the Berry brothers remained grounded, never forgetting their modest beginnings.

Dexter Mason Ferry and the D.M. Ferry Seed Company

The D.M. Ferry Seed Company, a horticultural giant with a history spanning over 150 years, stands as a cornerstone of the American seed industry. Founded by Dexter Mason Ferry in 1856 in Detroit, Michigan, this renowned company has made profound contributions to agriculture, gardening and the wider horticultural community.

Dexter Mason Ferry, a man of vision and agricultural acumen, established the company with the primary goal of providing high-quality vegetable and flower seeds to farmers and gardeners. His commitment to excellence and innovation set the foundation for what would become a trailblazing enterprise.

One of D.M. Ferry's earliest and most significant contributions was the introduction of new and improved seed varieties. Through meticulous breeding and selection, the company released numerous vegetable and flower cultivars that not only thrived in diverse climates but also consistently produced high yields. These innovations revolutionized the agricultural landscape, empowering farmers to grow more, with greater resilience and quality.

The company's commitment to research and development was exemplified by its early investment in experimental farms and testing grounds. These facilities allowed for rigorous testing of new seed varieties, ensuring they met the highest standards for germination, disease resistance and flavor. The resulting innovations reached far beyond Michigan's borders, benefiting farmers and gardeners across the United States and beyond.

D.M. Ferry's contributions extended beyond seed development. The company was a pioneer in packaging and marketing. Its distinctive red and white seed packets became iconic symbols of quality and reliability. Through clever branding and informative labels, they not only sold seeds but also educated and engaged their

Seed package (front cover), D.M. Ferry Seed Company. *Author's collection.*

customers. This approach was instrumental in making gardening accessible and appealing to a wider audience.

A significant milestone in the company's history was the publication of the *D.M. Ferry Seed Annual*, which provided detailed information on various seed varieties, along with practical advice for gardeners. The annual catalogue became a cherished resource for both novice and experienced gardeners, further solidifying the company's reputation as a trusted authority in horticulture.

D.M. Ferry also contributed to the horticultural industry by actively engaging with and supporting agricultural organizations and events. The company often participated in agricultural exhibitions and fairs, showcasing its latest innovations and providing educational resources to attendees. This participation not only promoted their products but also nurtured a sense of community within the industry.

The impact of D.M. Ferry Seed Company extended to the development of the seed industry itself. The company played a pivotal role in shaping industry standards and ethics. Their commitment to quality control, seed purity and accurate labeling set a high bar for the entire sector, inspiring others to follow suit. Dexter Mason Ferry's emphasis on integrity and transparency in business practices left a lasting legacy.

In the twentieth century, D.M. Ferry expanded its reach by merging with other seed companies, such as Burpee Seeds, strengthening its position as a leader in the industry. This strategic growth allowed them to offer an even wider selection of seeds to an ever-growing customer base.

Today, the D.M. Ferry Seed Company continues its legacy, adapting to modern agricultural practices while preserving its commitment to quality, innovation and service. Its influence on the agricultural and horticultural communities remains profound, with the company's name synonymous with reliability, trustworthiness and a rich tradition of excellence.

The D.M. Ferry Seed Company stands as a true giant in the American seed industry. Its contributions, spanning over a century and a half, have not only shaped the world of agriculture and gardening but have also left an indelible mark on the very fabric of the horticultural community. From its innovative seed varieties to its commitment to quality, packaging and marketing, D.M. Ferry's legacy continues to flourish, inspiring and guiding generations of farmers and gardeners.

Bernard Stroh and the Stroh Brewery Company: A Detroit Beer Legacy

The beer industry in Detroit owes much of its rich history and development to Bernard Stroh, a visionary entrepreneur who played a pivotal role in shaping the brewing landscape of the Motor City. Bernard Stroh's contributions to the beer industry in Detroit left an indelible mark, and the Stroh Brewery Company became a household name, synonymous with quality and tradition.

Born in Germany in 1822, Bernard Stroh immigrated to the United States in 1850, seeking new opportunities and a fresh start. Settling in Detroit, he quickly recognized the potential of the burgeoning beer market. Detroit's growing population and the influx of German immigrants created a growing demand for beer. This realization inspired Bernard Stroh to found the Lion's Head Brewery in 1850, marking the humble beginnings of the Stroh Brewing Company.

From the outset, Bernard Stroh focused on crafting beer with the utmost dedication to quality and authenticity. He adhered to the principles of *Reinheitsgebot*, the German Beer Purity Law, which stipulated that beer should be made from only water, malt, hops and yeast. This commitment to quality and traditional brewing methods set Stroh's beer apart in the Detroit market and became the foundation of the company's success.

The Stroh Brewery Company's growth was exponential, fueled by the increasing popularity of its lagers and ales. The company continued to expand its production capacity and distribution network, quickly becoming one of Detroit's leading beer producers. Bernard Stroh's emphasis on quality and innovation led the company to introduce refrigeration and pasteurization processes in its brewing, ensuring the freshness and consistency of their products.

One of Bernard Stroh's most significant contributions to the Detroit beer industry was his dedication to the local community. He believed in giving back to the city that had welcomed him as an immigrant. The Stroh Brewery Company became deeply involved in philanthropic efforts, supporting various charitable organizations and cultural institutions in Detroit. This commitment to community engagement and corporate social responsibility set a standard for other businesses in the city.

By the late nineteenth century, Stroh's beer had become a Detroit icon. The company's signature beer, Stroh's Bohemian-Style Pilsner, was widely recognized for its crisp, refreshing taste and became a favorite choice for

Stroh Brewery Company delivery truck, date unknown. *Author's collection.*

beer enthusiasts in the region. The distinctive Stroh's logo, featuring the iconic lion's head, further solidified the brand's identity and remained a symbol of quality.

Under Bernard Stroh's leadership, the Stroh Brewery Company expanded its operations to meet the growing demand, opening new breweries and acquiring other beer producers. This expansion not only strengthened the Stroh brand but also boosted Detroit's economy and employment opportunities. The company's workforce grew, providing jobs for many Detroit residents and contributing to the city's industrial prosperity.

While the Stroh Brewery Company had been a Detroit mainstay for generations, the brewing industry faced challenges in the late twentieth century. A series of acquisitions and mergers within the brewing industry led to a shifting landscape, and Stroh's, too, was affected. In 1985, the company was sold to Peter Stroh, Bernard Stroh's great-great-grandson, who tried to navigate the challenges of a rapidly changing industry.

Despite these challenges, the Stroh Brewery Company's legacy and Bernard Stroh's contributions to the Detroit beer industry remain vivid in the collective memory of the city. The commitment to quality, community and tradition that Bernard Stroh instilled in his brewery endures as a symbol of Detroit's brewing heritage. Stroh's beer remains a beloved choice among beer enthusiasts, and the Stroh name remains a testament to the enduring spirit of Detroit and the individuals who shaped the city's industries.

Bernard Stroh's contributions to the beer industry in Detroit are deeply significant. His pioneering spirit, commitment to quality and dedication to the local community established a brewing legacy that transcends generations. The Stroh Brewery Company's presence in Detroit symbolizes the city's rich brewing heritage, and Bernard Stroh's vision continues to inspire beer enthusiasts and entrepreneurs alike.

Thomas Witherell Palmer and Elizabeth "Lizzie" Palmer: Shapers of a Unique Legacy

In the annals of Detroit's history, certain names stand out as iconic figures who left an indelible mark on the city. Among these luminaries, Thomas Witherell Palmer and his wife, Elizabeth "Lizzie" Palmer, are celebrated as influential benefactors and visionaries who contributed significantly to the development of the city and the establishment of enduring institutions.

Thomas James Palmer was born in Detroit on January 25, 1830, into a family of prosperity and influence. His father was a successful dry goods merchant, and his mother was the daughter of Territorial Judge James Witherell. It was from this remarkable lineage that young Thomas inherited not only his family's legacy but also a thirst for knowledge and a sense of duty toward his community.

In his formative years, Thomas received a well-rounded education, which included stints in Detroit, St. Clair and a year spent at the University of Michigan. However, his desire for knowledge extended beyond the classroom. As a teenager, he embarked on an adventurous journey, touring Europe and South America, acquiring a worldly perspective that would shape his future endeavors.

At the age of twenty, in a symbolic act that demonstrated his reverence for his mother's heritage and his desire to honor her side of the family, he adopted his mother's maiden name, becoming Thomas Witherell Palmer. This choice was not only a reflection of familial devotion but also a declaration of his intention to carve a unique path in the world.

Thomas Witherell Palmer (1830–1913). *Library of Congress.*

In 1855, Thomas Witherell Palmer's life took a momentous turn when he married Elizabeth "Lizzie" Merrill. Lizzie was the daughter of Charles Merrill, Thomas's business partner, a noted figure with extensive holdings in the lumber industry. This union not only solidified business ties but also created a partnership rooted in shared values and a vision for a more equitable society.

Thomas began his career by involving himself in his family's insurance and real estate ventures, gradually building the foundation of his economic prowess. However, it was his collaboration with Charles Merrill in the lumber industry, specifically in St. Clair County and the Saginaw Valley, that propelled him to new heights. By 1863, he had ascended to the presidency of their lumber firm, further cementing his position as a prominent businessman and a key player in the region's industrial landscape.

Thomas Witherell Palmer was not merely a businessman. He was a fervent advocate for social and political causes that aligned with his vision of a more just and equitable society. As a Republican, he ardently supported the idea of federal control of the railroads. However, his most notable advocacy was for women's suffrage, a cause that would define his legacy.

His motto, "Equal rights for all, special privileges to none," echoed his commitment to fighting for the marginalized, be it women, children or animals. In 1877, he became the first president of the Michigan Society for the Prevention of Cruelty to Animals, which would later evolve into the Michigan Humane Society. His leadership and dedication to this cause showcased his compassionate nature and his unwavering commitment to making a difference.

Known for his engaging and persuasive oratory, Thomas Witherell Palmer's political career began to ascend. In 1878, he was elected to the Michigan State Senate, where he continued to champion the causes close to his heart. His influence continued to grow, culminating in his appointment as a United States senator, a position he held from 1883 to 1889.

Following his term as a U.S. senator, Thomas Witherell Palmer ventured into the realm of global diplomacy. He accepted an appointment as the U.S. ambassador to Spain, marking a significant chapter in his illustrious career. His diplomatic service would eventually pave the way for his role as the president of the 1893 World's Columbian Exposition Commission in Chicago, a prominent event of the era that celebrated the 400th anniversary of Christopher Columbus's arrival in the Americas.

In the grand tapestry of Detroit's history, Lizzie Palmer was not merely a supporting figure to her husband's endeavors; she was a formidable personality in her own right. Born in Lincoln, Maine, on October 8, 1837, she inherited her father's shares in the family business, an inheritance that would play a pivotal role in her philanthropic legacy.

Upon her father's passing, Lizzie commissioned the renowned architects Carrère and Hastings to design a fountain in his honor. Dedicated in 1901, the fountain initially graced the front of the old Detroit Opera House on Campus Martius. Later, in 1925, it was moved to Palmer Park, where it stands as a testament to her commitment to preserving her family's legacy and contributing to the city's cultural landscape.

Upon her death in 1916, Lizzie Palmer bequeathed $3 million to establish a school focused on educating young women about motherhood and family life. This generous funding laid the foundation for the Merrill-Palmer School, which eventually evolved into the Merrill Palmer Skillman Institute for Child and Family Development at Wayne State University. The institute remains at its original location, the Charles Lang Freer house, a testament to the enduring impact of her philanthropy.

In addition to their contributions to education, the Palmers were founding supporters of the Michigan Soldiers and Sailors Monument in Campus Martius, a monument that stands as a tribute to the veterans who served the nation. They were also instrumental in the establishment of the Detroit Museum of Art, which would evolve into the Detroit Institute of Arts, a cultural treasure in the city.

In their wills, the Palmers left a lasting legacy to Detroit. They provided funding for the Merrill Fountain, honoring Lizzie's father, and contributed to the establishment of the Mary Palmer Memorial Episcopal Methodist Church, a tribute to Thomas's mother. Their generosity extended to deeding 140 acres of farmland on the northern outskirts of Detroit to the city for use as a rural park. This property, complete with a log cabin, a small lake and a vast virgin forest, was soon rechristened as Palmer Park, a beloved and enduring recreational space in the heart of the city. Furthermore, a section of the Palmers' land would evolve into the Palmer Woods Historic District neighborhood, a testament to their influence in shaping the city's landscape.

The lives of Thomas Witherell Palmer and Elizabeth "Lizzie" Palmer are woven into the fabric of Detroit's history. Their contributions to business, politics, philanthropy and education have left an indelible mark on the city they loved. From advocating for women's suffrage and the welfare of animals to their generous gifts of parks, institutions and cultural treasures, the Palmers stand as exemplars of what a profound impact visionary individuals can have on a community. Their legacy endures, not just in the physical spaces that bear their name but in the lives and aspirations of countless individuals who continue to be inspired by their vision of a more just and equitable society.

James McMillan, John S. Newberry and the Michigan Car Company

Detroit's rich pre-automotive industrial history was shaped too by the rise of the Michigan Car Company, a key player in the late nineteenth century's railway car manufacturing industry. Founded during the Civil War in 1864 by James McMillan and John S. Newberry, the company embarked on its journey to provide rolling stock for the Union's war effort, a mission that would eventually lead to significant economic and technological contributions.

By 1873, the Michigan Car Company had established a sprawling thirty-acre manufacturing complex at the intersection of the Michigan Central and Grand Trunk Railroads in Springwells, just beyond the western city boundaries of Detroit. Initially, the company's rail cars were constructed from wood, but as technology progressed, steel emerged as the state-of-the-art material for production.

Fast-forward to the turn of the century, and the Michigan Car Company had solidified its place as Detroit's most critical manufacturing industry. They were producing an array of freight cars, including gondolas, tank cars, flat cars and hoppers, which were the workhorses responsible for transporting the nation's essential commodities like grain, oil, ore and timber.

The company's impact on the local economy was immense. During the mid-1870s, it boasted 881 production workers, with 566 directly employed by Michigan Car and an additional 315 by its subsidiary, the Detroit Car Wheel Company, along with 124 clerical employees. The wheel shop was a bustling center of activity, melting an average of fifty-five tons of iron daily to craft the massive thirty-three-inch-diameter, 570-pound car wheels. Skilled molders played a crucial role in pouring molten metal into wheel-shaped molds, and the wheels were then sent to the machine shop for finishing. Axles, provided by another subsidiary, the Baugh Steam Forge Company, were fitted together with the wheels, requiring intricate metalwork in the process. The blacksmith shop played a central role, and it was here that a young Henry Ford briefly found employment in 1879. A paint shop rounded out the operations at the Springwells site, highlighting the complexity of the production process.

The late nineteenth century was marked by economic fluctuations, with the Michigan Car Company experiencing alternating cycles of prosperity and hardship. In the early 1880s, the company generated substantial profits, reaching approximately $6 million annually, equivalent to a staggering $170 million today. However, the mid-1880s brought economic challenges, as harsh times took their toll. Workers of the Michigan Car Company played a prominent role in an unsuccessful citywide effort to institute the eight-hour workday in 1886. But as the economy rebounded, so did the company. By 1887, it had become one of the nation's most extensive railroad car shops, employing 2,500 workers capable of producing 10,000 cars and 110,000 car wheels annually.

The Michigan Car Company continued its expansion into the early 1890s, employing between three and four thousand workers in firms it either fully owned or held partial stakes in. This included the merger with the Peninsular Car Company in 1892, leading to the formation of the Michigan-Peninsular Car Company, which became the largest rolling stock manufacturer in the United States. The company employed five thousand workers and produced an impressive one hundred freight cars every day.

However, the economic rollercoaster took another dip when the national economic crash of 1893 hit, forcing Michigan-Peninsular to shut down for five months and lay off 1,200 workers. Freight car orders dwindled as over seventy railroad companies, including the Union Pacific, went bankrupt. The company's earnings plummeted from $867,000 in 1892 to $36,000 in 1894. Despite these challenges, the company proved resilient and rebounded by 1897, manufacturing forty-four thousand units.

In 1899, Michigan-Peninsular joined forces with a dozen other companies from around the nation to create the American Car and Foundry, a move aimed at consolidating railcar production and eliminating competition. Detroit's operations became the largest unit in the new company, producing about one-third of the company's output of 124,000 freight cars in 1899. By 1900, the company employed an impressive nine thousand Detroit residents, constituting about 7 percent of the city's industrial workforce. Business remained robust as the company's production reached an astounding 163,000 cars in 1902.

The year 1902 marked a significant turning point, with the introduction of an innovative production system in Detroit. This system involved using trucks to move rail cars under construction from one station to another, where specialized crews performed specific tasks before the car moved to the next stage. This concept of the moving assembly line, a precursor to the one famously introduced by Henry Ford in the following decade for automobile production, revolutionized the manufacturing process.

As time passed, the landscape of American Car and Foundry changed significantly. The original founders of the Michigan Car Company gradually withdrew from active involvement in the business, and the merged company established its headquarters in St. Louis. The production facilities at the Springwells site played vital roles during World War I, with some facilities adapted for munitions production. In the aftermath of the war, parts of the complex transitioned into motor coach manufacturing. Eventually, the site and its munitions plant were cleared to make way for the construction of the new Cadillac automobile production facility.

The Michigan Car Company and its descendants left an indelible mark on Detroit's industrial landscape, serving as a precursor to the city's automotive dominance and contributing significantly to its economic and technological advancements. Southwest Detroit, once home to bustling railway car factories, played a pivotal role in shaping the city's industrial heritage, ultimately paving the way for the rise of the automobile industry in the twentieth century.

Chapter 3

FROM OPEN-FIRE HEARTH TO CAST-IRON STOVE

The unearthing of vast deposits of copper and iron ore in the upper reaches of Michigan brought about an unprecedented wave of possibilities for enterprising individuals in Detroit. Among these visionaries, Jeremiah and James Dwyer stood out, uniquely poised to capitalize on these newfound natural riches.

While the Dwyer brothers were not the ones responsible for uncovering or transporting iron ore from the Upper Peninsula, nor did they lay claim to inventing the cast-iron stove, they were, in fact, beneficiaries of these pivotal discoveries and technological advancements. This fortuitous convergence of circumstances positioned them perfectly to spearhead a transformative industry in Detroit—the manufacturing of cast-iron stoves.

The cast-iron stove represented the latest advancement in a prevailing trend, especially in cold-weather regions, to discover more effective and efficient methods for both heating homes and preparing meals. This journey to harness the power of fire was marked by extensive experimentation and a considerable passage of time.

Fire as a Primal Force

The origins of the hearth and fireplace and cast-iron stove are deeply intertwined with the discovery of fire itself. For our ancestors, fire was not merely a tool but a transformative force that brought light, ancestors' warmth and protection from the darkness of the night. Its discovery is a defining moment in human history, and it reshaped the course of our evolution.

The earliest record of controlled fire dates back nearly one million years, with evidence suggesting that our hominin ancestors harnessed its power for warmth and protection. These early hearths were simple affairs, little more than a circle of stones to contain the flames. They provided not only warmth but also a sense of security and community, as early humans gathered around the fire to share food, stories and companionship.

The Birth of the Fireplace

As human societies progressed and settled into more permanent dwellings, the need for improved heating and cooking solutions became apparent. This led to the development of the first fireplaces, marking a significant step in the evolution of heating and cooking.

The ancient Greeks and Romans are credited with some of the earliest innovations in fireplace design. They introduced the concept of a central hearth or fire pit within the home, surrounded by a raised platform or bench. This setup improved ventilation and drew smoke upward, reducing indoor pollution. The Greeks also developed the idea of a chimney, albeit a rudimentary one, to further enhance smoke extraction.

In medieval Europe, fireplaces began to take on a more prominent role in the architecture of castles, monasteries and noble residences. These fireplaces featured intricate stone mantels and hoods designed to funnel smoke out of the buildings. They were often colossal structures, reflecting the grandeur of the structures they adorned.

During the Renaissance and Enlightenment periods, advancements in craftsmanship and design brought about more elegant and functional fireplaces. Elaborate mantels, ornate grates and decorative tiles became common features, showcasing the artistry of the time. Fireplaces were not only functional but also statements of wealth and refinement.

In the American colonies, hearths and fireplaces played an essential role in daily life. They provided warmth during harsh winters and served as the focal point for family gatherings. The hearth was not just a source of heat and light; it was a symbol of comfort, security and community.

Colonial fireplaces were often vast structures, capable of accommodating multiple pots and pans for cooking. Families would gather around the hearth

Early American hearth. *Library of Congress.*

to prepare meals, tell stories and find solace in the flickering flames. It was a place where both sustenance and cherished memories were created.

The hearth and fireplace, while essential, were far from perfect. They were inefficient, with a significant portion of the heat escaping up the chimney. The open nature of these traditional heating and cooking methods also posed significant fire hazards, and the abundant smoke they produced often led to poor indoor air quality. Moreover, the hearth and fireplace were labor-intensive, requiring constant attention to maintain a steady fire.

One very notable enhancement to fireplace design emerged during the late eighteenth century. The Rumford fireplace, designed by Sir Benjamin Thompson, also known as Count Rumford, addressed several shortcomings of the traditional fireplace. For example, traditional fireplaces were notorious for their inefficiency, as they allowed much of the heat to escape up the chimney. Rumford's design aimed to maximize the radiant heat output by reflecting the heat back into the room. The fireplace had a shallower firebox with angled sides that directed more heat into the room rather than up the chimney. Additionally, traditional fireplaces often suffered from poor smoke control, leading to smoky rooms. Rumford's design incorporated a streamlined throat and chimney, which enhanced the upward flow of smoke, reducing the chances of it billowing back into the room. Moreover, the Rumford fireplace's efficient design allowed for more complete combustion of wood, resulting in less fuel consumption compared to traditional fireplaces.

The Rumford fireplace's success and principles of improved efficiency had a lasting impact on fireplace design. The design was widely published and promoted, and many subsequent designs incorporated some of its features, gradually making fireplaces more efficient and less wasteful of heat, materials and fuel.

Early Experiments with Cast Iron: A Foundation for Innovation

The Industrial Revolution brought significant changes to the way people lived and worked. As factories and mills proliferated, people migrated from rural areas to cities in search of employment. Urbanization and the advent of new heating technologies began to diminish the importance of the hearth as the primary source of heat and cooking.

The quest for improvement in heating and cooking was perennial and included early experiments with cast iron, a durable and heat-retaining material that would play a pivotal role in the evolution of stoves. The cast-iron experiments began to take shape in the sixteenth century, introducing the world to the first semblances of what would eventually become the cast-iron stove. These rudimentary stoves were simple yet ingenious, typically composed of a cast-iron box with a fire chamber and a flue pipe that directed smoke away from the living space. The primary advantage of these early cast-iron stoves over traditional hearths and fireplaces was their improved heating efficiency. Cast iron's exceptional heat retention properties allowed these stoves to radiate warmth into the room instead of letting it vanish up the chimney.

However, these early stoves had their limitations. They were often cumbersome, complex to install and designed primarily for heating purposes. They lacked the precise temperature control necessary for versatile cooking. Yet they marked the first significant step toward revolutionizing indoor heating and cooking, and they paved the way for more sophisticated developments.

The Birth of the Cast-Iron Stove: Benjamin Franklin's Ingenious Contribution

As the eighteenth century dawned, a true luminary emerged in the realm of cast-iron stoves—Benjamin Franklin, an American polymath renowned for his contributions to various fields, including science, politics and, indeed, stove and fireplace design. It was in this latter endeavor that Franklin's genius truly shone, with the creation of what would become known as the Franklin stove.

Franklin's stove represented a pivotal moment in the evolution of heating and cooking technology. It addressed many of the shortcomings of earlier stoves and provided a blueprint for further innovations in cast-iron stove design. The advantages of Franklin's stove were manifold:

Improved Efficiency. Franklin's stove featured an innovative design with a hollow baffle system that prolonged the exposure of hot gases to the cast iron, significantly enhancing heat transfer and reducing fuel consumption.
Safety. By enclosing the fire within a cast-iron chamber, the Franklin stove significantly mitigated the risk of fire hazards compared to open hearths.
Better Temperature Control. The stove's design allowed for better temperature regulation through adjustments to the air intake and chimney damper, rendering it suitable for both heating and cooking.
Smoke Reduction. Franklin's stove effectively curtailed smoke emissions, thus enhancing indoor air quality and comfort.
Compact size. Unlike earlier stoves, Franklin's design was relatively compact, making it more amenable to installation within a wide array of homes.
Versatility. Its versatility made it an attractive option for both heating and cooking, ensuring it found a place in diverse households.

Despite its many advantages, Franklin's stove was not without its limitations. It primarily heated the immediate area around it, struggling to distribute heat evenly throughout larger rooms or multiple rooms. The stove's intricate design could be more costly to manufacture and maintain compared to simpler models. Additionally, like its predecessors, Franklin's stove still consumed significant amounts of fuel, often in the form of wood or coal, which could prove expensive.

Nevertheless, Benjamin Franklin's pioneering contributions marked a turning point in the journey of cast-iron stoves, inspiring further innovation and enhancement in heating and cooking technology. His legacy endures as a symbol of American ingenuity and the inexhaustible quest for comfort and efficiency.

The Gradual Transition from Wood to Coal

During the nineteenth century, obtaining firewood became a challenging task for both city dweller and country folk. Wood would remain the major source of fuel in the United States until 1890, when it was largely—but not exclusively—replaced by coal. However, the use of coal was making inroads by the 1860s in Detroit and elsewhere. Ready sources of firewood were becoming depleted, and the network of railroad tracks was making delivery of large quantities of coal from the East increasingly more feasible. The movement away from wood as a fuel was a calculated one and based on many factors.

- Availability of Timber: Deforestation and Land Clearance. In the nineteenth century, there was a significant demand for timber not only for firewood but also for construction and industrial purposes. This led to widespread deforestation in many areas, reducing the availability of nearby woodlands for firewood. Moreover, farmers in both urban and rural areas often cleared land for agriculture, further depleting local forests.
- Distance and Transportation. Firewood had to be transported from forests to urban and rural areas. In the nineteenth century, transportation was limited to horse-drawn carts, wagons or boats, which made hauling large quantities of firewood over long distances slow and expensive. Also, roads and transportation infrastructure were rudimentary, making it even more challenging to transport firewood efficiently.
- Competition for Resources. The growing industrialization of the United States during the nineteenth century increased the demand for wood as a fuel source for steam engines and factories. This competition made it harder for individuals to secure a sufficient supply of firewood. Furthermore, as cities grew, they required vast amounts of firewood for heating and cooking. The concentration of people in urban areas placed additional pressure on nearby wood resources.
- Property Rights and Regulations. Property rights and landownership often determined who had access to nearby woodlands. Regulations and

property disputes could limit access to firewood resources. In addition, in some regions there were early efforts to conserve forests due to concerns about deforestation and its long-term environmental impact. These efforts could restrict access to woodlands.

LABOR-INTENSIVE PROCESS OF HARVESTING AND PREPARING FIREWOOD. Cutting down trees, chopping wood and splitting it into manageable pieces required significant physical labor. This was a time-consuming and exhausting task for both city and country dwellers.

In the nineteenth century, cutting a cord of wood would typically take much longer than modern times due to the reliance on manual tools and less advanced equipment. Several factors would influence the time required:

TYPE OF WOOD. The type of wood being harvested played a significant role. Hardwoods like oak or hickory were denser and harder to work with, potentially adding more time to the process compared to softer woods like pine.

TOOLS AND EQUIPMENT. Woodcutters in the nineteenth century typically used hand tools, such as axes, handsaws and mauls, for splitting. The absence of chainsaws and hydraulic splitters would make the process slower and more physically demanding.

EXPERIENCE AND SKILL. The experience and skill level of the woodcutter were crucial. Experienced individuals who were skilled in using manual tools efficiently could work faster than those with limited experience.

TEAM SIZE. Having a team of people to work together could increase efficiency. One person could cut down trees, while others would be responsible for chopping, splitting and stacking the wood.

ENVIRONMENTAL CONDITIONS. Weather conditions, such as extreme heat, cold or precipitation, could affect the woodcutting process in the nineteenth century. Adverse weather conditions could slow down work and make it more challenging.

TRANSPORTATION. Depending on the proximity of the woodlands to the home, transporting the cut wood back to the house could be time-consuming, especially if it had to be done manually or with horse-drawn carts.

Given these factors and the limitations of nineteenth-century technology, it might take several days or even weeks for a small team of woodcutters to cut, split and stack a cord of wood. The exact time required would

Left: The housewife laboring over the hearth. *Right*: The modern housewife at her Jewel stove. *Author's collection.*

vary depending on local conditions, the size and density of the trees, the availability of labor and the efficiency of the tools used. The nineteenth-century woodcutting process was labor-intensive and physically demanding, and it required a combination of skill, experience and hard work. It was a critical task for many households during that era, as firewood was essential for heating, cooking and other daily needs.

Why Coal?

Coal fueled homes and factories until about 1940. The benefits of coal, once it became readily accessible, made firewood far less a reasonable option.

- Energy Density. Coal has a higher energy density compared to wood, meaning it provides more heat per unit of weight or volume. This increased energy efficiency made coal a more attractive option for heating homes. It allowed for smaller, more efficient stoves and furnaces that could heat a space effectively.
- Industrialization. The nineteenth century saw significant industrialization, which led to increased coal production. As coal mines became more productive, coal became more readily available and affordable. This made it a practical choice for both industrial and residential use.
- Transportation Infrastructure. Developments in transportation, such as the expansion of canals and railways, facilitated the distribution of coal from mining regions in urban centers like Detroit. This improved accessibility lowered transportation costs and made coal more widely available.
- Urbanization. As more people moved to cities and urban areas during the nineteenth century, the demand for heating fuel increased dramatically. Coal could be transported and stored more efficiently in urban environments, making it a practical choice for densely populated areas.
- Cleaner and More Predictable Burning. Coal, when burned in a controlled environment like a coal stove or furnace, produces a more consistent and controllable heat compared to wood. It also emits fewer sparks and less smoke, contributing to improved indoor air quality.
- Availability of Coal Stoves. The development and widespread adoption of coal-burning stoves and furnaces specifically designed for home heating further popularized coal as a heating fuel. These stoves were efficient and made it easier to regulate the heat in homes.
- Economic Considerations. In many regions, coal was cheaper than firewood due to its higher energy content and ease of transportation. The cost-effectiveness of coal made it an attractive option for households looking to reduce heating expenses.
- Longer Burn Times. Coal stoves could provide longer burn times compared to wood stoves, requiring less frequent refueling. This convenience was especially important during cold winter nights.

ENVIRONMENTAL IMPACT. Although the environmental impact of coal was not a significant concern in the nineteenth century, the shift to coal did contribute to the early stages of industrial pollution and smog in cities. However, people were initially more concerned with the immediate benefits of coal's heat output and convenience.

SHIFT IN FOREST CONSERVATION. As concerns about deforestation grew during the nineteenth century, there was increased pressure to conserve woodlands. This also played a role in the transition to coal as a heating fuel.

The decline of firewood as the primary home heating fuel before 1890 marked the end of an era in American domestic life. While firewood had been a reliable and abundant source of heat for centuries, changing societal dynamics, technological advancements and environmental concerns gradually led to its replacement by coal. The adoption of coal stoves and furnaces represented a significant leap forward in heating technology, offering greater efficiency and control.

Early Jewel stove with accessories. *Library of Congress.*

In retrospect, the transition from firewood to coal reflected broader shifts in American society, from agrarian to industrial, from rural to urban. It also underscored the importance of energy efficiency and economics in shaping fuel choices. While firewood continued to have a role in rural areas, the late nineteenth century saw the beginnings of a new era in home heating, setting the stage for further energy transitions in the decades that followed.

The Cast-Iron Stove Revolution

Cast-iron stoves revolutionized home heating and cooking in the nineteenth century, offering a level of versatility and efficiency that far surpassed traditional fireplaces and hearths. This transformation was driven by several key advantages that cast-iron stoves provided, making them a game-changer for households. Cast-iron stoves were a significant advancement over fireplaces and hearths due to their superior efficiency, even heat distribution, cooking capabilities, safety features, reduced maintenance, space efficiency and customization options. These stoves transformed the way people heated their homes and cooked their meals, greatly improving comfort and convenience in households during the nineteenth century. Their legacy continues to influence modern heating and cooking technology to this day.

Chapter 4

THE FOUNDING OF DETROIT STOVE WORKS

By 1860, Detroit's industrial landscape had reached a level of maturity that set the stage for the emergence of a transformative new industry, one that would not only provide livelihoods for thousands but also harness the rich iron ore reserves discovered in Michigan's Upper Peninsula. The city and region had developed the infrastructure, skilled workforce and economic stability necessary to accommodate this groundbreaking venture. It marked a pivotal moment in the region's history, as it prepared to tap into the abundant natural resources of the U.P., giving rise to a booming industry that would shape the economic trajectory of both Detroit and the entire southwest Michigan region.

JEREMIAH DWYER: THE FATHER OF THE DETROIT STOVE INDUSTRY

Jeremiah Dwyer, born in Brooklyn on August 22, 1837, came from a family with Irish roots. His father, Michael, had immigrated to America on a sailing ship and initially settled in Hartford, Connecticut. Michael spent eight years working on a Yankee farm up the Connecticut River. Eventually, he decided to move to Michigan with his young wife and their one-year-old son, Jeremiah, in search of a new life as farmers.

Initially, Michael's sights were set to the north of Detroit. He even hired Native American guides to explore the lake-rich Oakland County. However, his wife wisely suggested that they should find land closer to Detroit to access better markets for their farm produce. Heeding her advice, Michael purchased a 160-acre plot in the township of Springwells, about four miles west of Campus Martius, paying six dollars per acre. The Dwyers then began their hard work on the farm, and over time, two more children, a boy and a girl, joined their family.

Trademark registration, Jewel stove. *Library of Congress.*

The year 1848 marked a tragic turning point for the Dwyer family. While Michael was handling a team of spirited young horses near the railroad, they were suddenly frightened by the sight of a locomotive. The panicked horses bolted, causing chaos. In the midst of this turmoil, Dwyer was thrown from his seat on the buckboard and met a tragic and untimely end. This left behind his wife and their three young children.

Following the loss of his father, Jeremiah, who was just eleven years old at the time, made a valiant effort to support his mother in running their farm. However, it quickly became apparent that their efforts were not producing the desired outcomes. Recognizing the necessity of affording her children better educational opportunities than what their rural surroundings could offer, Jeremiah's mother made a momentous decision. She sold their countryside property for ten dollars per acre and acquired a residence in the city of Detroit. In this new urban setting, Jeremiah and his younger siblings gained access to a few years of education in public schools.

From an early age, Jeremiah harbored the aspiration of working in an iron foundry. During his school years, he would often pass by Barclay's foundry and be captivated, as well as perplexed, by the way the cold, solid iron he understood so well could flow like "cherry-colored molten water." At that time, securing an apprenticeship and mastering a trade was a formidable challenge. Nonetheless, due to financial constraints, Jeremiah knew he needed to find employment. He managed to secure a job at the sawmill operated by Smith and Dwight, where he remained employed for approximately a year. Eventually, through his network of friends, Jeremiah successfully secured an apprenticeship in the art of molding at the Hydraulic Iron Works, owned by the firm of Kellogg and Van Skoye. However, the terms of his apprenticeship—essentially an indenture agreement—were quite demanding. He was committed to serving four years and making up for any time lost. As he read through the imposing, meticulously worded document, he realized that to fulfill these obligations, he would have to forsake many of the indulgences and distractions of youth. There was no room for gambling, drinking or undesirable companions. He resolved to live a disciplined life, getting as much sleep as time would allow. The days would be long, but he was determined to master every facet of this labor-intensive trade.

In his initial year, he earned a meager $2.50 per week, working ten hours per day. His pay increased to $3.50 in the second year, $4.50 in the third year and finally $6.00 in the fourth year. To the satisfaction of his employers, he dutifully fulfilled these obligations, receiving a glowing letter of recommendation upon completing his apprenticeship, a testament to his unwavering dedication and exceptional skill. He would safeguard that letter, as well as the indenture contract, until around 1871, when they were misplaced. In the years that followed, he often mentioned that he would offer $500.00 to anyone who could recover those documents, for they represented more to him than any of his material possessions—a tangible reminder of the practical education and unwavering work ethic that had shaped his life.

Following his apprenticeship, Jeremiah honed his craft as a journeyman in stove foundries across Buffalo, Rochester and Troy, New York. Eventually, he returned to Detroit. However, due to health concerns arising from prolonged exposure to the trade's toxic environment, he temporarily accepted a position with the D&M Railroad, where he worked for about a year. Later, an enticing offer came his way to work as foreman at one of Detroit's renowned foundries, Geary and Russell.

In 1861, an agricultural reaper and stove foundry venture kicked off at the intersection of Mount Elliot Avenue and Wight Street, thanks to the efforts of Messrs. Ganson and Mizner. Unfortunately, this enterprise encountered significant manufacturing challenges and eventually folded. The property fell into the hands of T.W. Mizner, who approached Jeremiah with a proposal to create a new stove business in partnership. Jeremiah had $3,000 saved to invest from savings during recent years. Following negotiations, the company called J. Dwyer & Co. was founded, and while small, the firm proved profitable during its first two years. During this time, Jeremiah designed,

BROILED BEEF STEAK.—Have the steak cut from one inch to one and a half thick. If there is any danger of the edges curling, cut across the gristle several times. Have the oven heated at least five minutes before broiling. Arrange the steak compactly upon a small pan just large enough to hold it; this will economize the drippings and prevent their taking fire. Place the steak about an inch below the burners in the horizontal broiler of a Detroit Jewel, and when brown on one side turn it and brown the other; or suspend it in the broiling rack of a Detroit Jewel that has a perpendicular broiler, which will broil it on both sides at once. In the latter case the broiling may begin about as soon as burners are lighted. Avoid piercing the meat as this would cause an escape of the juice, but thrust the fork into the edge where the gristle offers a firm support. Press on the meat to ascertain when it is done; as soon as the elasticity gives way it is rare in the middle and "done to a turn." Dish on a hot platter, and dress with salt, pepper, and butter and the addition of any garnish which taste may suggest. The time required for broiling in the horizontal broiler will be from ten to fifteen minutes, according to the thickness of the steak. Much less time is required for broiling in our perpendicular broiler, which broils both sides at once.

Broil chops in the same way.

MEATS AND FISH

Recipes for meat and fish on a Jewel stove. *Author's collection.*

produced and marketed a range of cast-iron stoves under the brand name Defiance. The initial design was a simple four-hole wood burner.

Many years later, after he had left to establish the Michigan Stove Company, Jeremiah sought to locate one of his original Defiance stoves. He intended to include it in his exhibit at the Columbian Exposition in Chicago in 1893, where he aimed to showcase the significant advancements in stove manufacturing in Detroit over three decades. Serendipitously, one day, a German immigrant arrived at the factory with an original 1862 Defiance stove in her farmer's wagon. The lady had brought the stove in for a minor hinge repair.

Jeremiah was thrilled to see one of his original creations and invited the elderly lady into the showroom. He offered to exchange any new stove on display for the old Defiance. Her initial response was negative, as she exclaimed, "*Nein, nein,* this stove is just fine. It's the best stove. I want no better stove." Nevertheless, after a salesman on the floor welcomed her

Baking recipes for use on a Jewel stove. *Author's collection.*

and showed her some of the latest models, she relented. She considered a particular model that could burn both wood and coal, which she believed would suit her needs. She planned to use wood in the winter and switch to coal as required.

Thus, Jeremiah achieved his goal of locating a Defiance stove and was able to demonstrate, at the 1893 Columbian Exposition, the remarkable progress made in the stove industry in Detroit over thirty years of design and manufacturing history.

W.H. Tefft later acquired Mizner's share in the business, yet the firm retained its original name, J. Dwyer & Co., for an additional year. Then, in 1864, Merrill I. Mills, who had recently become prosperous from the manufacture of tobacco, joined forces with them, and together, they founded a stock company called the Detroit Stove Works, with Jeremiah assuming the role of manager. This expansion brought about significant enhancements in the company's design, manufacturing and marketing capabilities.

Regrettably, while overseeing the construction of a new manufacturing facility in Hamtramck in 1869, Jeremiah's health problems, stemming from his deep involvement in all aspects of stove production, resurfaced, necessitating his departure from the foundry once more. Following advice from his personal physician, Jeremiah embarked on a journey to the South for an indefinite period of rehabilitation. However, before his departure, he negotiated the sale of his stake in the Detroit Stove Works to his brother, James.

After just a couple years of rehabilitation, Jeremiah was anxious to resume the only line of work he felt both comfortable with and competent in—stove making. In 1871, Jeremiah returned to Detroit and organized the Michigan Stove Company, with financing provided by investors Charles DuCharme, Richard Long, Merrill I. Mills and George H. Barbour, all of whom assumed administrative positions within the new company.

Avoiding as much as possible the toxic environmental hazards that previously sent him to the South, Jeremiah hired and trained workers committed to producing a wide range of quality stoves—fueled by wood, coal, oil and eventually gas. To expand the market for Michigan Stove Company products, Jeremiah engaged in a series of clever marketing campaigns designed to put his stoves front and center in the consumers' minds. The first included creating a giant stove replica of the company's new flagship brand, the Garland, for the 1893 Columbian World's Fair to be held in Chicago. Later, in 1900, he would take several models of his company stoves to Paris for its International World's Fair in 1900, during which his Garland models would be awarded a first-place designation in the stove division of domestic technology.

Throughout his extensive career, Jeremiah faced a multitude of challenges, ranging from employee injuries, both fatal and nonfatal, to a catastrophic fire in 1907 that nearly destroyed the company's primary facility on Jefferson Avenue. He managed to successfully navigate the shifting tides of consumer preferences and grappled with the complexities of expanding into foreign markets.

Jeremiah's dedication to addressing employee injuries was unwavering. Having toiled at every workstation during his apprenticeship and journeyman years, he gained an unparalleled firsthand understanding of the unique hazards associated with working with molten iron. This knowledge encompassed the entire process, from melting pig iron to meticulously polishing the final pieces and preparing them for assembly. In a workplace where foremen were notorious for their stern approach, often resorting to

Cooking by Gas booklet distributed by Detroit Stove Works. *Author's collection.*

or threatening to use disciplinary measures on younger, slower workers, Jeremiah emerged as a beacon of empathy and concern. Given that the majority of iron foundry workers—in Detroit and elsewhere—were typically between sixteen and twenty years old (with some as young as thirteen but few as old as fifty), Jeremiah was keen on engaging with them not only about production efficiency but health and safety issues as well. If someone fell ill or was absent due to sickness, he personally checked on them, driven by the belief that the well-being of his workers was inextricably linked to the overall success of their labor.

Changing consumer preferences was a perennial issue affecting the company's bottom line. In a 1906 interview with the *Detroit Free Press*, Jeremiah explained it this way:

> *There is an immense amount of detail, and one style follows another with rapidity. I often compare it with the millinery business. Patterns that sell well today have little or no demand next season; and there is endless rivalry in bringing out new models.*
>
> *A stove that will sell in one part of the country will not go in another part. In various cities, for example, gas pressure differs materially, and our designers must take that into consideration. In still other places, cities are built on the side of a hill, and when the wind is in a certain adverse direction, naturally the draft is affected. Stoves, to be right, must have unusually good draft. In other parts, the fuel is wood; in others, hard coal; in still others, soft coal, and vice versa. There are innumerable conditions known to the trade which make the stove business one calling for an intimate knowledge of the geography of the country.*

In addition to regional differences, individual taste played a considerable role in the consumer market. Again, Jeremiah explains:

> *We decorate with nickel, spun brass, aluminum, tiles, knobs, buttons, and other ornaments; vary the size, style and appearance, and so exacting have people become that we often hear a request for a stove without a draft or without a damper. In fact, people expect more and more of a stove—requiring it to work automatically, I may say, without the slightest attention. An amusing instance came to my wife's notice some time ago. She was calling on a lady who had used one of the coal stoves for six years. "Do you know," said the good lady, "the stove worked well for six years, but now refuses to burn. When the draft is open it is all right, but when it's closed, the fire dies right down." "Have you cleaned the flues?" asked Mrs. Dwyer. "The flues, the flues; what are they?" I sent a man around to see, and he found the flues choked with the accumulation of six years.*

Yet another challenge facing the stove industry was the demand of foreign trade. Jeremiah, or his representative, traveled the world to establish markets for stoves manufactured by the Michigan Stove Company. He learned during one trip that the Japanese had no use for stoves. Cooking was done in a stone pot, fueled by a bit of charcoal in an open flame, which boiled the rice. In France, Germany and other European countries, the cost of American stoves was considered too expensive, and Europeans needed to be educated about the quality of the higher-priced American goods.

By contrast, Jeremiah had fundamentally good things to say about the American consumer, despite personal taste and regional differences. Americans, he noted, always bought the best products within their budgets. And despite the challenges created by segmented markets, Jeremiah remained optimistic about the future of the stove industry for the rest of his life. Barring some major and unforeseen economic downturn, Jeremiah believed the industry could easily double during the quarter century ahead.

One of the most formidable challenges that Jeremiah encountered during his illustrious career as the "father of the Detroit stove industry" occurred in January 1907, when a devastating fire threatened to consume the entire Michigan Stove Company. The inferno raged uncontrollably for nearly three hours, engulfing various departments and resulting in the tragic death of one bystander, severe injuries to at least six individuals and the displacement of over two thousand employees.

Despite the company being underinsured, facing significant financial risk, the most critical components of the plant, situated within the original company building along Jefferson Avenue near Adair Street, remarkably remained unscathed. Essential records, invaluable patterns and stoves that were meticulously prepared for shipment miraculously escaped the flames. The company displayed remarkable resilience, swiftly embarking on a rebuilding process, and within a few months, it was back to full operational capacity.

On January 30, 1920, Jeremiah Dwyer passed away at the age of eighty-three. It was said that there was never a worker strike at Michigan Stove Company because of the profound respect plant employees had for him. Learning on one occasion of some employee dissatisfaction over compensation, Jeremiah is said to have given every plant employee an immediate 10 percent raise. Pallbearers at his funeral were men from the plant floor of the Michigan Stove Company. Jeremiah was laid to rest at Mount Elliot Cemetery, just a short ten-minute walk from the stove company that he founded.

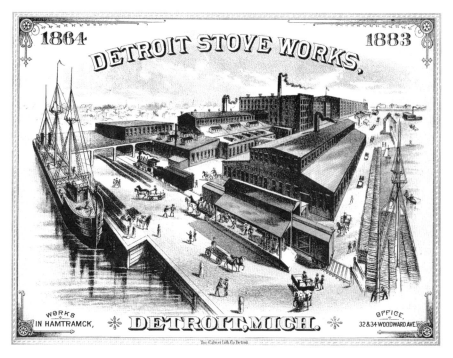

Detroit Stove Works celebrating twenty years of continuous operation. At this time, the Hamtramck plant extended all the way to the Detroit River. *Author's collection.*

The Detroit Stove Works: A Success Story

In 1861, Detroit Stove Works had its humble beginnings as a foundry, marking the inception of this pioneering industrial venture in the Midwest. Four years later, in 1864, a stock company purchased the establishment, and the reborn enterprise, keeping the name Detroit Stove Works, was officially incorporated with an initial capital of $50,000. As time rolled on, the company's financial strength saw remarkable growth. By 1865, its capital had swelled to $100,000 and subsequently surged to an impressive $300,000. This metamorphosis paved the way for the Detroit Stove Works to evolve into one of Detroit's most substantial industrial giants. With its sprawling ten-acre campus, stretching from Jefferson Avenue to the river, the stove-manufacturing company provided gainful employment for approximately 1,400 individuals as it melted sixty tons of iron daily.

Situated strategically, the works was well connected for both water and rail transportation. Adjacent to the river, they offered easy access to waterways, and through the Transit and Belt Line Railways, they established links with all the major railroad networks converging on Detroit. The convenience was underscored by the Transit Railway, which had its terminus within the works' yards, and the Belt Line Railway terminating merely two blocks away. The physical footprint of Detroit Stove Works encompassed 325,000 square feet, an area that saw steady expansion as new buildings sprang up over time to accommodate the ever-growing demands of the business.

Much of the company's direct engagement with dealers took place at its Chicago branch, strategically located on South Canal Street. The eastern U.S. market was catered to by a branch in Buffalo, New York. But the reach of Detroit Stove Works was not limited to American shores; it boasted several European agencies, prominent among them being Frankfort, Paris, London, Brussels and Vienna. The company had also nurtured a thriving export trade, making its mark in far-flung destinations such as South America, Australia and Tasmania.

The products of Detroit Stove Works sold under the brand Jewel stoves and ranges, available in a staggering array of over eight hundred different sizes and styles, catering to every conceivable type of fuel. These marvels of modern stove-making seamlessly incorporated the finest features known to the industry. Subject to annual remodeling, they adapted to the ever-evolving demands of the market, a gold standard in both stove construction and design. With an annual sale of more than sixty thousand units, the popularity of the Jewels bore testimony to their exceptional quality and the widespread appreciation they garnered from the public.

The Dwyer brothers were keenly attuned to the history and marketing of cast-iron stoves along the East Coast. Their attention couldn't help but be drawn to the well-respected Jewett brand, produced by Jewett and Root in Buffalo, New York, and named after its principal owner, Serman Skinner Jewett. Their brand had made waves in the market, and with branches in Detroit, Chicago, Milwaukee and San Francisco, it's highly likely the Dwyer brothers were familiar with its success.

Yet the Dwyers had a different vision. Rather than simply affixing their family name to their cast-iron products, they sought to create a distinct identity that would not only capture the imaginations of their customers but also serve as a rigorous internal standard. They recognized that in a fiercely competitive market, their product had to be consistently flawless, akin to a meticulously crafted diamond.

As they brainstormed, the word "jewel" entered their thoughts. It could have been a subtle nod to the Jewett name, or perhaps it was a sudden, brilliant stroke of inspiration that led them to this elegant and evocative moniker.

The Process of Stove Making

Crafting stoves was a meticulous art, and while the process could vary over time and across different manufacturing facilities, the steps outlined here generally align with the techniques employed by major stove manufacturers in Detroit during the nineteenth and early twentieth centuries.

Step One: Stove design is a collaborative effort between the designer and management. They assess various factors such as market demand, special orders, target demographics, seasonal considerations and more to determine the type, size and shape of stoves to produce.

Step Two: The stove designer closely collaborates with a draftsman, unless they happen to be the same person. The abstract stove concept takes its first concrete form on paper, eventually evolving into a detailed working drawing. This drawing encompasses every aspect, including the ornamental details on the front and sides, providing a comprehensive blueprint for the casting process.

Step Three: The patternmaker assumes a crucial role in the process. With precision and care, he crafts an exact wooden replica of the drawings in relief. In some cases, a molding clay reproduction of the drawing is created to offer a tangible preview of the final product. While wooden patterns are suitable for creating iron working patterns, they aren't practical for actual casting due to their susceptibility to damage. Therefore, wooden patterns are used exclusively for crafting ironworking patterns. Several working patterns are made to meet the demand of the iron foundry, but significant work is required before they can be used. The iron pattern for each of the stove's components—its body, doors, legs and other decorative elements—must be flawless in terms of detail and craftsmanship. Each undergoes meticulous hand-finishing, including polishing to a gleaming finish, and the elimination of any rough edges or corners. Once perfected, the pattern is ready for use and is handed over to the foundry workers for their daily tasks.

Step Four: The mold-maker now must work very quickly as the molten iron is simultaneously being prepared for pouring. A mold is made by packing special molding sand, typically a mixture of silica sand and a binder such as clay, tightly around the pattern. The mold is made in two halves—the drag (bottom) and the cope (top)—which are separated later to remove the pattern and pour the molten iron. After the sand is packed around the pattern, the pattern is removed, leaving behind a cavity in the shape of the stove's component. This cavity is where the molten iron will be poured. In the case of complex stoves with internal spaces, sand cores may be placed within the mold to create hollow sections. Cores are made separately and positioned inside the mold before closing it. The drag and cope are brought together, aligning the two halves precisely to form a complete mold. The mold is typically held together with pins and clamps.

STEP FIVE: Inside the foundry, multiple cupola furnaces serve a unified purpose: to generate the cast iron necessary to meet the day's production demands. The cupola furnace is a tall, vertical structure lined with refractory materials. It has a cylindrical shape with tuyeres (nozzles) at the bottom for introducing air. To start the process, the cupola is charged with a mixture of pig iron ingots and other iron scrap, such as cast iron or steel scrap. The scrap is loaded from the top. A controlled ignition source is used to start a fire at the base of the cupola. This fire creates the initial heat required to melt the pig iron and scrap. Once the fire is established, a powerful air blower is used to introduce a high-velocity stream of air into the cupola through the tuyeres. This air supply is essential for combustion and for maintaining a high temperature within the furnace. As the air is introduced into the cupola, it creates a high-temperature environment within the furnace. The pig iron and scrap metal gradually melt due to the intense heat. In this superheated environment, the carbon in the pig iron combines with oxygen from the air to form carbon monoxide (CO). This exothermic reaction produces even more heat, further raising the temperature within the cupola. The molten iron, with its impurities and slag (waste materials), sinks to the bottom of the cupola because it is denser than the hot gases and slag. The impurities and slag form a layer on top of the molten iron. The liquid iron is periodically tapped from the bottom of the cupola, while the slag is skimmed off the surface. When the molten iron reaches the desired composition and temperature, it is tapped from the cupola through a tap hole. The iron is collected in a ladle or mold and is now referred to as "cast iron."

STEP SIX: With the molds in place and prepared, the molten iron from the cupola is transferred to a vessel known as a ladle. The ladle is used to transport the molten iron to the molds. It is essential to maintain the molten iron's high temperature and prevent premature solidification. In addition to the molds, a gating and runner system is usually incorporated. This system includes channels and gates that direct the molten iron from the ladle into the mold cavity. It helps ensure even distribution of the molten iron and minimizes turbulence during pouring. The operator carefully pours the molten iron into the molds in a specific sequence to avoid defects. The sequence often begins with the smaller or more intricate sections of the mold, allowing the molten iron to fill the entire mold progressively. Once the molten iron is in the mold, it starts to cool and solidify. The rate of solidification depends on factors such as the iron's temperature, the mold material and the size and thickness of the cast-iron part being created. After the cast iron has sufficiently cooled and solidified, the molds are typically cooled further, often with water or air, to speed up the process. Once solidification is complete, the molds are opened, and the cast-iron parts are removed from the mold cavity. This removal process is called "shakeout." The cast-iron parts may require additional finishing processes, such as grinding, machining, cleaning or heat treatment, to achieve the desired quality and appearance. The cast-iron parts are inspected for defects or imperfections, and any necessary quality checks are performed. Defective parts may be discarded or repaired. In some cases, multiple cast-iron parts may need to be assembled together to create the final product. This assembly step is carried out as part of the manufacturing process. The process of transferring molten iron into molds requires precision and control to ensure that the cast-iron parts are of high quality and meet the intended specifications.

Pig iron is an intermediate product in the process of making steel, and it's an essential raw material in the iron and steel industry. It is named "pig iron" because it was traditionally cast into molds that resembled piglets, which gave it its name. Pig iron is produced from iron ore, coke (a type of coal) and limestone in a smelting furnace, typically a blast furnace.

In the nineteenth century, the process was highly labor-intensive, with skilled craftsmen playing a crucial role in every step of production. This artisanal approach resulted in uniquely designed and crafted stoves, each with its own character and quality. The stoves produced during this era are still, to this day, admired for their craftsmanship and historical significance.

James Dwyer, who had been with Detroit Stove Works since the very beginning, decided it was time to venture forth and establish a new stove company, so on March 23, 1891, Dwyer, along with investors William B. Moran, William R. Harmount, Robert M. Campau and Christian H. Buhl, organized Peninsular Stove Company.

Detroit Stove Works Merges with Michigan Stove Company

In 1923, Detroit Stove Works acquired the Art Stove Company, which was situated in Milwaukee Junction. Since Detroit Stove Works had no need for the smaller property, they decided to sell the headquarters and showroom to the Frank L. Bromley Company, an industrial real estate broker. Later in the 1960s, the nearby factory and warehouse were demolished to make way for the Chrysler Freeway. Interestingly, the original headquarters and showroom, now empty, remain as the sole surviving structure that bears witness to Detroit's once-thriving stove industry.

With the merger of Detroit Stove Works and the Michigan Stove Company in 1925, the newly christened Detroit-Michigan Stove Company expanded its operations beyond manufacturing stoves and furnaces. It began producing gas ranges for household use and heavy-duty heating and cooking appliances for hotels, clubs, restaurants and institutions, marketed under the Garland and Laurel brands, as well as Jewel and Detroit Jewel names. Notably, in 1927, it showcased the impressive Garland stove in a grassy area adjacent to the Detroit-Michigan Stove Company near the entrance to Belle Isle.

While fiscal year 1926 marked a successful period for Detroit-Michigan Stove, with net sales totaling $8.1 million and net profits reaching $1.2 million, the subsequent decade, marked by the Great Depression, saw a decline in annual net sales. Sales plummeted to as low as $2.3 million, and the company faced losses each year from 1931 through 1934, as well as in 1938. However, in fiscal year 1940, the firm managed to turn the tide, posting a net profit of $210,000 on net sales of $3.1 million, which allowed them to initiate dividend payments. Notably, it phased out furnaces in the mid-1930s.

In 1945, Detroit-Michigan Stove significantly boosted its revenues by acquiring A-B Stoves, Inc. of Battle Creek, Michigan. Additionally, it expanded into metal fabrication, producing components for automotive and other manufacturers. The peak of its success was reached in 1948, when net sales soared to $21 million and net profits neared $2 million. However, in 1949, sales plummeted by nearly half and failed to recover significantly. The company suffered losses exceeding $1 million in 1953 and over $1.6 million in 1954, despite generating sales of only $9 million.

In 1955, when Welbilt Stove acquired Detroit-Michigan Stove Company, the merged entity became Welbilt Corp., a publicly traded corporation that retained Detroit-Michigan Stove Company's listing on the New York Stock Exchange. Sadly, in 1957, the Detroit plant closed permanently, marking the end of an era for the company.

Chapter 5

THE MICHIGAN STOVE COMPANY

In 1869, Jeremiah Dwyer oversaw the construction of the new Detroit Stove Works facility in Hamtramck. However, in the winter of 1870, due to overexertion and exposure while relocating and starting up the new works, he fell seriously ill with a severe cold that settled in his lungs, complicating an already existing heart problem. Upon his physician's advice, he journeyed south. Fearing he might not return, he sold his interest in the business to his brother James.

After only a short period in the South, Jeremiah grew restless and returned home in the summer of 1871, encouraged by Alfred and Charles DuCharme to reenter the stove-manufacturing industry. His initial plan was to settle all his business in town and then move to Virginia, where he would establish a new stove-making business. However, his old business friends pressured him to reestablish himself in Detroit, where he had family, friends and numerous business connections. So in the fall of 1871, he partnered with Charles DuCharme and Richard H. Long, and they acquired the Ogden & Russel property at the foot of Adair Street, near the outlet of Bloody Run. They immediately set about gathering the necessary materials for a new stove manufacturing facility.

However, due to an early onset of winter, their construction plans were delayed. During the winter months of 1871–72, M.I. Mills proposed adding his property, located along Jefferson Avenue and Adair Street, at the initial cost. He also expressed his desire to join them in this enterprise and offered his property to partner with the others. The group accepted his offer, and

Michigan Stove Company employees, date unknown. *Author's collection.*

a few months later, they were joined by George H. Barbour. Together, they established the Michigan Stove Company, with Charles DuCharme as president, M.I. Mills as vice president, George H. Barbour as secretary, R.H. Long as superintendent and Jeremiah Dwyer as manager. As spring approached, they accelerated the construction of their building at the corner of Jefferson Avenue and Adair Street. The manufacture of stoves for sale and distribution did not technically begin until September 12, but once the first wave of stoves was produced, the company proved to be an immediate success. In 1873, the company's first full year of operation, 8,825 stoves were manufactured and shipped out. Ten years later, in 1883, 52,338 stoves were built. The plant was processing over 17,000 tons of pig iron a year, and that number was climbing. Over the years, the facility would continue to expand operations to accommodate its growing domestic and international clientele.

The factory buildings, from 1022 to 1052 Jefferson Avenue, were 360 by 700 feet in dimension, the grounds constituting an area of over sixteen acres. The facility would offer a complete line of cooking and heating stoves and ranges, made under one name, one trademark—the Garland— and one equal and uniform grade, averaging from two hundred to three

GARLAND STOVES AND RANGES

Left: Trademark registration, Garland stove. *Right*: Trademark registration, Garland stove, January 20, 1883. *Library of Congress.*

hundred stoves per day and from sixty thousand to seventy thousand yearly. Even early on, employment would range from between 1,000 and 1,200 workers. Later, the number would grow to over 2,200.

The material used was from a high-quality grade of iron ore from Lake Superior; Hanging Rock, Ohio; and the Chattanooga, Tennessee and Birmingham, Alabama iron mines. A specialty was made of aluminum mixed with cast iron for the production of Garland stoves and ranges, the only line in the world made from this valuable combination. Employment of aluminum in this combination with cast iron produced smooth castings, prevented cracking and added additional strength.

Large branch offices for the sale of the Garland stoves and ranges were established in Chicago, Buffalo, New York City and several foreign cities.

GEORGE H. BARBOUR: MASTER MARKETER

Jeremiah would one day reflect on his life during a newspaper interview and freely admit that there were many people responsible for his success. However, if there was one person at Michigan Stove Company who produced the greatest impact on the company's bottom line, beyond Jeremiah himself, it might well be George H. Barbour.

George H. Barbour proved to be the ideal business partner for Jeremiah Dwyer at the Michigan Stove Company. While Dwyer excelled in the manufacturing aspect of the business, Barbour brought his expertise to the sales and marketing side. Despite the generally acknowledged high quality of their cast-iron stoves, it was clear that these products wouldn't sell themselves. The competition from New York was fierce, and they were not willing to yield any ground to the budding Detroit companies, regardless of the product's quality.

Barbour possessed a unique talent for garnering attention for the stoves produced in the Detroit facility through ingenious marketing strategies and by effectively training field salespeople. His reputation extended far and wide, earning him the respect of peers nationwide. He not only secured the position of president in the National Manufacturers Association but also found himself on the boards of several banks and businesses.

However, George H. Barbour's enduring legacy in the realm of Detroit business revolves around his ingenious idea: creating a colossal replica of a Garland stove for display at the Columbian Exhibition in Chicago in 1893. This innovation would forever define his impact on Detroit. Less well known, but no less important, was his commitment to creating a permanent historical tablet memorializing the "Bloody Run" episode in Detroit history on the grounds of the Michigan Stove Company, very near where the historical event actually occurred in 1763.

Garland stove accessories on display at Michigan Stove Company store. *Library of Congress.*

George Barbour was born in Collinsville, Connecticut, on June 26, 1843. Starting at nine years old, Barbour divided his day between learning from books at the local public school and learning the art of salesmanship from his father in the family's general store. He always attributed the most crucial lessons he learned about sales and marketing to those early days, when he interacted with a diverse range of customers in a store that offered a wide variety of dry goods and groceries.

When George turned eighteen and learned of his father's desire to retire, he engaged in negotiations for a half-interest in the business, with the payment spread over time. In the following year, upon his father's complete retirement, George's father sold his own half-interest to his brother-in-law, J.E. Goodman. The store was then rebranded as Goodman and Barbour, and George brought his new partner up to speed. This partnership proved to be a success, enabling him to pay off his debt to his father within just two years. Just a year later, he took over J.E. Goodman's share and successfully ran the store as the sole owner for the next seven years.

At the age of twenty-nine, George Barbour found himself yearning for a fresh and more significant challenge. It wasn't long before the perfect opportunity presented itself. His brother, Edwin S. Barbour, who happened to have married the daughter of William Tefft, a co-founder of Detroit Stove Works and who brought him into the company, extended an invitation for George to join forces with the other partners, Jeremiah Dwyer, M.I. Mills and Charles DuCharme.

Recognizing his brother's unwavering work ethic and impressive track record in sales, Edwin believed that George was the ideal addition to the budding stove industry in Detroit. As a result, George made the bold decision to sell his general store, using the proceeds to negotiate a partner's stake in the company. He also took on the role of secretary and manager of the sales department with great enthusiasm.

Once he had established himself in Detroit, George Barbour discovered that Jeremiah Dwyer's philosophy of offering high-quality products at reasonable prices resonated with his own principles. Harkening back to his days at the general store, he consistently adhered to this philosophy by stocking only the finest merchandise and selling it at a fair price, ensuring a reasonable profit margin. It was through this straightforward approach that he cultivated a loyal customer base.

In 1872, when he assumed his new role, Barbour swiftly realized that his previous experience, where local patrons frequented his store regularly, wouldn't suffice. He now needed a strategy to tap into new external markets and dealers. Initially, he collaborated with an existing salesperson at the Michigan Stove Company to develop the local market, specifically targeting small towns across Michigan. Expanding beyond the existing territory, especially toward the competitive eastern markets, was imperative.

One anecdote illustrates the pivotal moment of Barbour's expansion efforts. During a pitch to a dealer, a Michigan Stove Company salesman opened a hatbox, revealing a finely crafted, nickel-plated spun brass top that was intended to adorn the company's new baseburner. The brilliance of the top not only left the dealer in awe but also led a rival stove manufacturer's salesman to question his loyalty. In fact, this encounter prompted the salesman to realize, "The moment I saw that top, I understood that traditional sales methods were obsolete, and my future lay with this innovative company." He promptly wrote two letters, one tendering his resignation from his current employment and the other seeking a position with the Michigan Stove Company.

Among the many highlights of George Barbour's tenure at Michigan Stove Company were overseeing the design, creation and transportation of the gigantic Garland stove for the 1893 World's Columbian Exposition held in Chicago to celebrate the 400th anniversary of Christopher Columbus's arrival in the New World in 1492; being awarded first place among domestic stove entries for his Garland brand stove at the Exposition Universelle (or World's Fair) in Paris in 1900; and addressing the tragic fire that nearly destroyed the entire manufacturing plant in January 1907.

The Giant Garland Stove at the Columbian National Exposition of 1893

In 1890, an exceptional opportunity arose to showcase the Garland brand of the Michigan Stove Company. This opportunity coincided with the planning of the Columbian Exposition, also known as Chicago's World's Fair, which aimed to celebrate the 400th anniversary of Christopher Columbus's arrival in the New World in 1492.

Thomas W. Palmer, a prominent figure from Detroit who had previously served as both a Michigan state and United States senator, had been appointed as the ambassador to Spain by President Benjamin Harrison on March 12, 1889. Subsequently, Palmer was appointed to the National Commission of the Columbian National Exposition, where he was elected as the commission's president by his fellow commissioners.

Postcard depiction of Michigan Stove Company's Chicago World's Fair Exhibit, 1893. *Author's collection.*

Each state was tasked with sending two representatives to participate in the organizing activities of the Columbian Exposition. Michigan had designated M. Henry Lane and Charles H. Richmond for this role. However, when Richmond resigned his appointment, George H. Barbour, the alternate, stepped in to take his place.

It was at this juncture that Barbour recognized a remarkable marketing opportunity to promote his company's brand. Rather than simply showcasing the products of the Michigan Stove Company, Barbour envisioned something much more dramatic: a colossal replica of their classic Garland stove. He shared this vision with Jeremiah Dwyer, and together, they turned this concept into a reality.

Talented woodcarvers were commissioned to create the gigantic replica, which weighed a whopping fifteen tons and stood at impressive dimensions of twenty-five by thirty by twenty feet. Transporting this massive replica to Chicago required the use of three freight cars, and it was eventually displayed in the Manufacturer's and Liberal Arts Building at the Columbian Exposition.

The giant Garland stove quickly became a favorite among the fair's attendees, drawing large crowds and earning a place of prominence alongside

Postcard panoramic view of the Exposition Universelle, Paris, 1900. The Eiffel Tower can be seen in the distance. *Author's collection.*

other notable attractions, such as the enormous Ferris wheel in Jackson Park and the Krug gun. After a successful six-month run, the giant Garland was disassembled and transported back to the Michigan Stove Company. It was then put on display just inside the wrought-iron fence, becoming a local and tourist destination for many years.

Exposition Universelle of 1900

The remarkable marketing success of the colossal Garland stove, evident through its immense popularity at the Columbian Exposition in Chicago and its subsequent status as a tourist attraction upon installation on Jefferson Avenue, instilled in George Barbour a profound sense of accomplishment. The prospect of promoting the Garland brand was a constant undercurrent in Barbour's thoughts, and in 1899, another promising opportunity materialized. Paris, France, was slated to host yet another world's fair, the Exposition Universelle of 1900, intended to herald the dawn of a new millennium. More than seventy-five thousand businesses were set to showcase

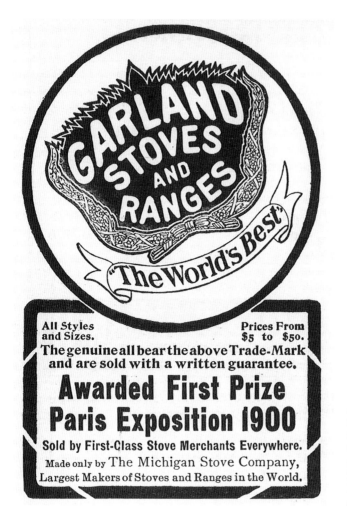

A widely distributed newspaper advertisement for the Garland stove after being awarded first prize, Paris Exposition, 1900. *Author's collection.*

their latest technological marvels, including innovations like electric fire engines, dry cell batteries, escalators and talking films.

The exhibition's thematic pavilions organized French and foreign exhibitors by product categories. Barbour seized this opportunity to ship several different models of Garland stoves to Paris, each showcasing the most recent advancements in cooking and heating, utilizing various types of fuels. His products earned the prestigious title of first place in the domestic technology category, as determined by the fair's judges. This triumphant outcome upon his return bolstered Barbour's confidence that the Michigan Stove Company was poised to lead the charge into the new millennium as the foremost player in its field.

An Outrageous Tragedy

Given the intense heat generated by the three cupolas in the iron foundry department and the scorching two-thousand-degree temperature of the molten iron used in the sandcasting process, it's truly remarkable that the Michigan Stove Company managed to evade any major disasters in its thirty-five-year history. However, on the fateful evening of Tuesday, January 8, 1907, the entire stove manufacturing facility came perilously close to being reduced to ashes, saved only by a twist of fate in the form of shifting winds and the swift response of the fire department.

At approximately 6:30 p.m., Morgan O'Keefe, the vigilant night watchman, spotted flames and billowing smoke emerging from the coal-hopper in the engine room. O'Keefe acted swiftly, sending out an urgent distress signal to the factory's private service, which in turn promptly alerted the fire department. It's worth noting that in 1907, there was no direct communication link between the factory and the fire department.

Over the next three hours, the fire spread relentlessly from the epicenter of the building, which rapidly became a blazing inferno. Trucks numbered 6, 7 and 17 were dispatched to the scene, soon joined by additional firefighting resources. Their efforts to contain the spreading blaze were impeded by a malfunctioning fire hydrant that burst when they attempted to connect to it. The fire propagated through different departments, including molding, mounting, polishing, nickel-plating, shipping and patternmaking.

Intermittent explosions of flammable materials served as ominous reminders of the unpredictable and challenging nature of the fire. Walls crumbled and fell in rapid succession, posing a severe threat. Just before 8:00 p.m., the roof of the expansive storage building, housing around thirty thousand stoves, was torn asunder with a deafening roar and a resounding crash.

In the initial hours of the fire, the imposing Garland stove, prominently displayed just within the cast-iron fence along Jefferson Avenue, stood as a symbol of defiance. However, its very survival now hung in the balance as the relentless flames continued their advance.

At 8:50 p.m., an alarming event unfolded near the massive Garland stove at the plant's closest wall. Firefighters dashed for safety as the solid brick structure of the wall gave way, toppling to the ground and taking a portion of the stove's supporting pedestal down with it. This left the stove's legs buried beneath the rubble.

Shortly after 9:00 p.m., it seemed that the fire was finally coming under control. Remarkably, the original 1872 building's main offices, including the front display room, remained unscathed. The rear section of the foundry also appeared to have escaped the blaze's reach.

The company's ongoing expansion necessitated enlarging the plant. Even prior to the fire, construction was in progress on a seven-story, 150-by-60-foot addition and a new foundry building near the Detroit River. These additions would almost double the plant's capacity, demanding an acceleration of their completion.

Regrettably, the fire resulted in at least six hospitalizations, all of whom were bystanders. Tragically, one person succumbed to their injuries at St. Mary's Hospital, while the remaining six suffered injuries ranging from broken bones to severe lacerations. Fortunately, there were no burn victims.

During the fire's outbreak, C.A. DuCharme, the secretary-treasurer, and George T. Barbour, the general manager, observed the destruction from one of the windows in the original 1872 office building. Meanwhile, Jeremiah Dwyer, the company's president and founder, stood with thousands of onlookers on the sidewalk, witnessing the years of hard work go up in smoke.

The cost of demolishing the fire-ravaged structures and constructing new facilities to replace the destroyed departments was, at that moment,

The aftermath of the Michigan Stove Company fire. The giant Garland stove collapses into debris. *Walter P. Reuther Library, Archives of Labor and Urban Affairs, Wayne State University.*

purely speculative. George Barbour, however, immediately recognized that the expense to the company would far exceed the $400,000 in property insurance held.

The following morning, a group of several hundred men, primarily consisting of company employees, embarked on the challenging task of clearing the extensive debris left in the aftermath. Meanwhile, inside the corporate offices, George Barbour was devising a plan to quickly restart business operations. Construction work on the two new structures was already well underway, and the on-site construction company was soon to be tasked with expediting the process.

Simultaneously, corporate executives assessed the extent of the damage and were relieved to find that the flames had miraculously spared the plant's most crucial components. Among the departments that emerged unscathed from the fire were the foundry, engine rooms, boiler room with its two substantial vertical boilers, mounting room, castings storage department and castings warehouse. Notably, the display room, with its valuable samples and a substantial quantity of finished stoves ready for shipment, was also preserved. To be extra cautious, important records and books were relocated during the fire. Among the most crucial items saved was the collection of patterns. The more valuable patterns were stored in a fireproof vault, while those in the workshop remained undamaged. The company quickly arranged to rent a small frame building adjacent to the eastern property, where pattern work could be reestablished on a smaller scale without delay.

In just a matter of days, company officials confidently announced that plant operations would soon resume. The company delivered on its promise. A mere nine weeks after the day much of the plant was reduced to ashes, an army of 750 construction workers rushed the completion of the two new buildings. These structures, housing various departments and a fresh foundry, were constructed from concrete, brick and steel beams supported by cast-iron columns. With the sudden addition of 252,000 square feet of floor space, the company was now fully equipped to re-commence the production of Garland stoves. There was little reason to doubt that the number of manufactured Garland stoves in this, the year of the massive fire, would compete favorably with the previous year's record production of ninety-five thousand stoves.

In 1925, after much time and thoughtful family consideration, Michigan Stove Company officials entered into formal merger discussions with their closely connected rival, Detroit Stove Works. The terms of the merger would require shutting down the existing facility at Jefferson and Adair and moving

Pouring molten metal into sand molds at Michigan Stove Company. *Walter P. Reuther Library, Archives of Labor and Urban Affairs, Wayne State University.*

its now joint operation farther east, near the west entrance ramp to Belle Isle. The president of the new company, Detroit-Michigan Stove Company, William T. Barbour, nephew of George H. Barbour, a founding member of Michigan Stove Company, would take the Bloody Run tablet into his personal possession for safekeeping until the Big Stove was moved and both could be rededicated at the new site.

Chapter 6

JAMES DWYER AND THE PENINSULAR STOVE COMPANY

James Dwyer had already proven his mettle as an exceptional manager and later as a superintendent at the Detroit Stove Works. Yet after fifteen years, he hungered for a fresh challenge. Perhaps it was spurred by the success of his older brother, who had founded the thriving Michigan Stove Company from scratch. Aware of the market's capacity for another stove manufacturer, Dwyer was confident that entering this arena could be successful if the quality of its products matched or exceeded their competitors.

In the early spring of 1881, Dwyer convened with three potential investors—William B. Moran, William R. Hermont and Robert Campau. They swiftly embraced Dwyer's vision for a new line of stoves. By late March, the quartet had reached a tentative agreement to establish a new enterprise, even without a definitive name for the proposed company. Just three weeks later, on Saturday, March 17, 1881, these investors officially founded the Peninsular Stove Company by filing articles of incorporation.

Construction of the new facility, designed by Mason and Rice, commenced on August 29. Remarkably, a mere six months later, on February 18, 1882, under the meticulous supervision of James Dwyer, the Peninsular Stove Company initiated the production and distribution of its inaugural lines of stoves for consumers from its new facility at Fort Street and Trumbull Avenue.

Trademark registration, Peninsular Stove Company. *Library of Congress.*

During the six-month period of building construction, Dwyer, working with close associates, began designing a broad range of cast-iron products and establishing the requisite regional and international network of dealers for a seamless launch once the facility doors swung open. So well prepared in this new venture was James Dwyer that by the end of the following year, 1883, Peninsular Stove company could boast the manufacture and distribution of over 20,000 units. By the end of the decade of the 1880s, Peninsular would offer consumers a choice of more than 270 different models of its top-tier cast-iron stoves.

Who Was James Dwyer?

Like his older brother Jeremiah, James Dwyer faced the harsh realities of life from a tender age. The loss of his father when he was only six, followed by the passing of his mother at the age of twelve, left him with no choice but to seek guidance and support from his older sibling. However, Jeremiah was already fully employed at that point, making James's journey even more challenging.

Following his mother's death, James embarked on an apprenticeship with Charles Kellogg and Company to learn the trade of a machinist. For five years, he honed his skills while working diligently. His ambition led him to New York, where he spent an additional six years as a journeyman, perfecting his machinist trade. At the age of twenty-four, James finally felt ready to make his way back to Detroit and join his brother at the newly established Detroit Stove Works. Together, they worked tirelessly to develop a process unmatched in the industry for the quality of the stoves it produced.

It was in 1869 that James assumed the role of general manager, as his older brother had to step down due to health issues. Over the span of the 1870s, James remained at the helm of Detroit Stove Works, steering it through remarkable growth and establishing it as the foremost stove making enterprise in the entire region. However, James observed that the industry held a wealth of untapped potential, ripe for innovation and the introduction of fresh approaches and brands. From his perspective of more than a decade in the industry, James could see an exciting landscape where new ideas and distinctive brands could flourish and carve out their own niche in the market.

During his more than two decades' tenure at the helm of the Peninsular Stove Company, Dwyer would see tremendous growth for his company, one

PENINSULAR STOVES AND RANGES

Work Better

Bake Quicker

Last Longer

CONSUME LESS FUEL

...AND...

Give Greater Satisfaction

THAN ANY OTHER MAKE.

A WRITTEN GUARANTEE

ACCOMPANIES EACH.

Left: A widely circulated newspaper advertisement for Peninsular stoves. *Author's collection.*

Below: Peninsular Stove Company postcard. *Author's collection.*

day becoming the leader of the second-largest stove-manufacturing business in the country. The disappointments he would encounter included a brief labor-management dispute, a building fire and a building collapse.

Though James suffered no reported health issues—but did survive a train car accident in Ontario where the engineer was killed—he passed away unexpectedly, but peacefully, from "old age" as one obituary suggested, at the age of sixty-nine at his summer cottage in Grosse Pointe. At the time of his death, his three sons all held positions within Peninsular Stove Company—James M., Edwin L. and Albert E.

In the late nineteenth century, the three stove companies operated by the Dwyer family—Detroit Stove Works, Michigan Stove Company and Peninsular Stove Company—largely managed to avoid labor disputes. As Thomas Klug points out, "Among the largest, most combative, and best organized of the city's trade unions was Iron Molders' Union No. 31, a union representing the city's stove molders. During the 1890s, the international molders' union worked out a system of nation-wide collective bargaining with an association of leading stove manufacturers, a system that served as a model for other metal-working trades."

The Labor Strike of 1887: Skilled Tradesmen Rebel Against the "Buck System"

Despite the generally peaceful labor climate, an exception occurred on June 4, 1887. On this date, around five hundred stove molders employed by the Peninsula Stove Company and the Michigan Stove Company walked out on strike. This strike was the result of unsuccessful negotiations with both companies concerning the implementation of a new labor management system known as the "buck system," which had been recently introduced.

The final decision to go on strike was not reached capriciously. Within the iron molders' union, officers and skilled tradesmen debated the issue for a full day and a half. The union members simply concluded that any establishment of a buck system would be "ruinous," and the general feeling was that they would never submit to it.

The buck or helper system was the practice by the company of apprenticing to each molder a boy (around age fourteen) who helped the molder in his work. The disadvantage to the molders in this system was that after two or three years, if any significant conflict arose between the molders and the employer, the latter could lock them out and retain the "bucks," who would by that time be pretty well versed in the trade. An agreement had been entered into, according to the molders, that no helpers should be forced on them after May 1. The two companies, however, were reluctant to abandon the system, while the molders were demanding the complete abolition of the system.

The strike did not affect the Detroit Stove Works, as no helpers were employed by this, the original stove company in Detroit. Relations between the company and the molders were said to be peaceful and harmonious.

At Michigan Stove Company, George Barbour expressed surprise at the strike actions of the molders. Speaking on behalf of management of both Michigan and Peninsular Stove Companies, Barbour advanced two reasons for a continuation of the system, at least in some form. First, Barbour argued, no one on the floor was being forced to accept a helper. It was strictly voluntary, as agreed upon. To abolish a voluntary system was, in Barbour's mind, un-American and a "kind of tyranny" that the company would never submit to. The second reason to continue the buck system, according to Barbour, was for the benefit of the older worker. For example, both Michigan and Peninsular Stove Companies employed molders as old as fifty, who, in most instances, had been performing their work admirably. However, now, after thirty years of strenuous and dangerous sand casting, the older, still highly skilled molder might not have the endurance he once did and could clearly benefit from the added assistance of an unskilled buck. If this older employee had a helper to do some of the heavy lifting, as it were, he could continue to compete with younger skilled tradesmen and make the same competitive wages. Without the assistance of helpers, many of the older workers' jobs could easily be threatened by competitive younger tradesmen. Ultimately, both sides agreed to a compromise that the buck system would exist only as a voluntary and not a prescribed activity.

THE TRAGIC FIRE OF 1893

On Tuesday evening, November 1, 1893, a fire broke out in the nickel-plating and polishing department, located on the top floor of Peninsular Stove Company's four-story building facing out onto River Street. During daytime hours, as many as two hundred employees would have been involved in various jobs within the department, but at this time of night, only a few employees were present and were somehow able to escape the blaze. Unfortunately, four of the firemen responding to the fire were not as lucky, and several of the responders had narrow escapes as the floor was completely gutted by flames. During the progress of the fire, fireman David Boyd had his right arm literally ripped off at the elbow. The line of hose used to put out the fire was controlled by a small motor situated outside on River Street near the entrance to the building. Boyd was on one side of the engine, and a second fireman, John Dalley, was on the other side. The signal had been given to stop the engine while the line of hose supplied by it was being shifted around to eliminate kinks and put it in a more favorable position. Boyd took advantage of the momentary pause to oil up the machine. He had his right arm through the flywheel when the signal came for water, and Dalley, unaware of what Boyd was doing, started the engine from the other side.

In a moment, the flywheel began whirling, and Boyd's arm was severed clean from his body. He fell to the ground screaming, in obvious agony and losing a great deal of blood. Several who witnessed the accident rushed forward to help him. A Grace Hospital ambulance was sent for. Remarkably, Boyd rose from the ground with the assistance of two police officers who had arrived at the scene just in time to witness the accident. Boyd looked at his shredded rubber coat sleeve in disbelief, under which he was continuing to bleed profusely.

Boyd walked with the police officers three blocks to the corner of Seventh and Fort Streets, arriving there just as the ambulance arrived. He was rushed to St. Mary's Hospital, where the emergency surgeon found it necessary to amputate slightly more of his arm. Fortunately, Boyd survived the ordeal. Three other firemen suffered minor injuries, mainly from falling debris.

The total damage to the building was estimated at $45,000 in 1893 dollars, two-thirds of which was covered by insurance.

The fire should have been a warning to each of the three stove-making establishments because, as company officers—and the general public through news reports—would learn years later following another building disaster, steps both by the companies themselves and by city building inspectors had not been taken to ensure the safety and welfare of their employees. In the case of both the Michigan Stove Company fire in 1907 and this much smaller blaze at one of the Peninsular Stove Company's buildings in 1893, the loss of life was limited only by the time of day, when most employees were home with their families. On another day, at another time, the loss of life could have been substantial.

1909: Another Tragic Fire

In 1909, two years after the disastrous Michigan Stove Company fire, another disaster occurred, this time at Peninsular, and only fortuitous circumstance prevented a significant loss of human life. As it was, one employee was killed, and two others were injured. At 10:40 a.m. on Saturday, June 12, 1909, a large section of the sixth floor of Peninsular Stove Company's main office building collapsed without warning, setting off a chain reaction. Floor after floor collapsed—or, as one eyewitness described it, "broken like egg shells"—and the chaotic mass of timbers, stoves, crates and stove parts crashed through to the basement. Wreckage was piled as high as the second floor. The basement and first floor were full, and the second floor was strewn with broken stoves and timbers.

Fortunately, few workers were present, most of the main office workforce taking the usual Saturday holiday. Only four men were carried down in the debris. Two of these were rescued by men who placed their own lives in peril from the possible further collapse of adjoining walls. The two men were severely injured. Paul Szurek, a laborer, was so badly injured that surgeons at Red Cross Hospital held out little hope for his recovery. The other, Thomas Egan, a shipping clerk, was also severely injured but was expected to recover. A third man, Thomas Stephenson, the warehouse foreman, suffered three broken ribs and an injured shoulder. The fourth man, William Hollar, traffic manager for the company, lay buried under the mass of wreckage. It's hard to imagine that he could have escaped his fate that particular day. His office was on the main floor, toward the Fort Street side of the building. He went up on the third floor to see about the shipping of some stoves. He was likely standing right in the middle of the floor when the crash occurred. There was little to no chance for him to escape the devastation that followed.

John M. Dwyer, secretary of the company and son of the founder, immediately inquired of all staff on duty. "I am satisfied there are no others in the wreckage," he told a reporter for the *Detroit Free Press*. "We have taken a careful inventory of our men and we are sure now that poor Hollar is the only victim." On every one of the six floors, stoves were piled rows deep. The building contained practically the whole stock of the company to be sold for the winter of 1909–10. A gas pipe in the basement burst, starting a fire that crept steadily through the ruins. The building was never in danger from the blaze, which was extinguished in about half an hour, but there was serious concern that any workmen trapped in the wreck—notwithstanding Secretary Dwyer's assessment—might be cremated before they could be rescued.

Hollar's body was found twenty-seven hours after the building's inner collapse. The body was in a half-standing, half-reclining position. The fingers on both hands were torn, as though he struggled to free himself. The autopsy surgeon would later conclude that it was unlikely that Hollar was killed instantly. The autopsy revealed evidence of suffocation. Both of his legs were fractured at the ankles, three ribs were broken and his body was covered with bruises and lacerations.

The cause of the collapse, according to the coroner who thoroughly inspected the premises in the days following the death of William Hollar, was likely that one or more floors were overloaded with stoves and stove parts, beyond the physical weight the joists could support. Especially troubling to Coroner Burgess, however, were reports reaching him that a portion of one of the floors came near collapsing four years earlier. As James Dwyer was in Europe on business, his son John, acting manager of the plant, denied knowing anything about such a report. Even more troubling, however, was that there was no record in the office of the building commission of the Peninsular Stove Company ever having been inspected. "If this is true of the Peninsular plant," the coroner suggested, "how many more factory buildings in Detroit are in the same category."

The editorial page of the *Detroit Times* on June 14 was awash with critical questions and commentary about worker safety and the fate of William Hollar's widow and the five children left behind after the tragedy. An editorial opined:

> *An answer is due from the coroner's office and the public wants it for the sake of hundreds of thousands of men, women and children in this city who have to work for a living in buildings provided for by them. Only one life was lost, a rather remarkable fact, but one which does not lessen the seriousness of the proposition....He was engaged in his regular duties taking him here and there, from floor to floor, when, in the twinkling of an eye, he was sent down beneath the mass of wreckage and pinned fast. How long was it before death came to relieve him as he lay possibly conscious of his fate? Can you imagine the agony he suffered; the thoughts of loved ones at home, his helplessness; then probably loss of reason and finally, but mercifully, death. If there is a pen in any hand that could exaggerate the horror of the accident so far as William Hollar was concerned, though it told of a hundred killed? And such a fate for William Hollar alone*

demands that no more working men in the city of Detroit be held subject to a like ending of their days, that no more widows be made by the falling of floors under which their husbands work, and that no more children be robbed of fathers toiling to feed and clothe the bodies of children.

In the hearts of Detroit residents, these editorial sentiments were steadily gaining momentum. As the city's workforce entered the twentieth century, a gradual but profound change was unfolding in Detroit and across the United States. This shift brought to the forefront pressing societal concerns, including the perils of unsafe working conditions, the exploitation of child labor and the advocacy for workers' rights. Upton Sinclair's novel *The Jungle* in 1906 shed light on the deplorable conditions in the meatpacking industry, sparking outrage and reform efforts. Just a few years later, the tragic Shirtwaist Factory fire in 1911 in New York's garment district, in which 146 garment workers—mostly immigrant women—were killed, underscored the dire need for workplace safety regulations.

A 1912 delivery truck for Peninsular Stove Company. *Burton Historical Collection, Detroit Public Library.*

In Detroit, on the same day in 1907 when thousands of residents witnessed the devastating fire at the Michigan Stove Company, readers of Detroit's major daily newspapers could not have missed the front-page stories out of Pittsburgh and the Jones and Laughlin Steel Works. Even as the last cinders were being extinguished at the Michigan Stove Company, an explosion of gas at the base of a furnace sent tons of molten metal around the furnace for a radius of forty feet. Twenty-four workers were literally evaporated by the molten torrent; seven others were rushed to the hospital with severe burns.

Workaday Hazards, Accidents and Injuries

Before 1971, when the State of Wisconsin enacted the first comprehensive and compulsory workers' compensation law, injured workers—or their survivors—had to resort to common law remedies. To be awarded compensation for loss or injury, they often had to prove negligence or wrongdoing, which could be a lengthy and challenging legal process. Employers generally had the upper hand in legal disputes.

In 1912, at the same time that a comprehensive workers' compensation law passed into law in Michigan, Jeremiah Dwyer, president of the Michigan Stove Company, became the first executive of a major company to contract with an insurance company to provide each of his employees with life insurance through a "group policy," a new insurance product; only one other, Montgomery Ward and Company in Chicago, had negotiated such a policy for its employees. Michigan Stove Company would absorb the cost of several thousand dollars. Under the plan, each married man was insured for $500, while each single man or boy was insured for $250, the policy being payable immediately upon death. Detroit Stove and Peninsular would soon follow suit.

Workers at Peninsular found little solace in the fact that job-related fatalities could provide some temporary support for their families. Their more immediate concern revolved around their daily exposure to physical hazards and a toxic work environment whose long-term medical consequences remained largely uncharted. While the peril of handling molten iron was widely acknowledged at a rudimentary level, the numerous associated risks confounded the limited medical knowledge of the early twentieth century, rendering them virtually untreatable.

It wasn't until 1922 that the federal government took its initial steps toward addressing these pressing concerns, with the establishment of the Bureau of Labor Standards under the Department of Labor. A significant stride was made in 1970, when the Occupational Safety and Health Administration was formed, ushering in a more comprehensive approach to safeguarding worker health and safety.

There is little in the way of statistics on accidents—in Detroit or nationwide—before the early 1900s, in part because there existed a mistaken perception of safety that was held by both employees and employers. Specifically, it was believed that each worker was individually responsible for their own safety, with skilled craftsmen accustomed to self-management and personal safety responsibility. The idea that it was the management's duty to provide a safe work environment was not considered. The mindset was influenced by the earlier craft production era, which was suitable for situations where workers primarily used hand tools, but it became inadequate in the context of industrial factories where one worker's actions could potentially pose risks to others.

For cast-iron stove tradesmen in the late nineteenth and early twentieth centuries, the workday hazards were numerous, a few of which included:

DUST. In a casting foundry like Peninsular Stove Company, dust is formed throughout most stages of the casting process. Most of the dust is generated during the process of making molds and cores. The dust composition would depend to some degree on the binder used to keep the sand tightly packed. Removing the mold after the cast iron has cooled typically generates a lot of dry sand, which was much more toxic than wet sand. If the cast-iron foundry was exposed to high concentrations of silica, workers could develop lung-related diseases such as bronchitis, tuberculosis, silicosis and other complications. With long-term exposure, there is a risk of lung cancer.

BODILY INJURY ACCIDENTS. (a) *Eye Injuries.* Lack of protective eyewear of facial shields exposed workers to embedded silica from sand casting, splashed from molten metals and irritation from caustic chemical exposure; (b) *Musculoskeletal disorder.* Ergonomic solutions were not available in such metal molding and processing facilities to alleviate the constant physical toll of lifting heavy ladles of molten iron, bending, working in awkward body postures and performing the same or similar tasks repetitively; and (c) *Infections.* Untreated cuts, bruises, lacerations and burns were frequently ignored and further aggravated under pressures imposed by production deadlines, leading to bacterial infections.

Explosions. Explosions in the iron foundry cupolas typically occurred due to the buildup of volatile gases, such as carbon monoxide, within the furnace. These gases accumulate when there is incomplete combustion of the fuel used for heating the furnace, leading to a flammable mixture. When this mixture reaches a critical concentration and comes in contact with an ignition source, such as a spark or an open flame, it can result in a sudden release of energy, causing an explosion. Proper ventilation and safety measures are essential to prevent such incidents.

Noxious Gases and Vapors. Gases and vapors in an iron foundry can be dangerous to health because they may contain toxic substances like carbon monoxide and particulate matter, which can lead to respiratory problems, as well as irritants and carcinogens that can harm the lungs and other organs. Additionally, exposure to excessive heat and humidity in foundry environments can cause heat stress and related health issues.

Seldom recognized as environmental factors, the oversight in inspecting facilities for structural deficiencies and fire hazards emerged as critical safety risks. These oversights resulted in catastrophic incidents at both the Michigan Stove Company and the Peninsular Stove Company during the late nineteenth and early twentieth centuries.

In the mid-1920s, officials at Peninsular sought to upgrade their production methods, mirroring the efficient models prevalent in Detroit's automobile factories. The existing manufacturing site sprawled across more than seven acres at Fort and Trumbull, but its segmented layout hindered modern production practices.

There were persistent rumors that the Pennsylvania Railroad aimed to acquire this property for a huge passenger terminal. Meanwhile, Peninsular was exploring various properties across different parts of Detroit, each with railway access. Behind closed doors, Peninsular officials kept negotiations discreet with one potential suitor. Contrary to expectations, they engaged in talks with the Michigan Central Railway, surprising many as Pennsylvania Railroad had never made a bid for the property. Michigan Central's interest lay in utilizing the space for yard and storage warehouse facilities.

Throughout the summer of 1926, negotiations continued until Peninsular set a date, September 14, for a stockholder meeting to vote on the proposed sale to Michigan Central. Speculations by real estate experts estimated the sale might reach up to $5 million due to the property's size and prime downtown location.

Even while negotiating with Michigan Central, Peninsular started experimenting with their new venture: large-scale manufacturing of oil-burning stoves and furnaces. They planned to construct a one-story facility embracing assembly-line techniques in a semi-automated setting, aiming for a 50 percent efficiency boost.

On September 14, the stockholders, many descendants of Jeremiah and James Dwyer, approved the sale of the Fort Street property to Michigan Central Railroad for $2.5 million. Peninsular intended to maintain a branch office downtown in a smaller building just north of the former site on Fort Street.

The Move to Brightmoor and Subsequent Bankruptcy

Fast-forward to May 7, 1927, a mere eight months after shareholder approval of the downtown factory sale: Fred T. Moran, Peninsular president, took a symbolic shovel from his five-year-old grandson and broke ground on the new facility. The location, at Burt Road and the Pere Marquette Railroad in the Brightmoor area of northwest Detroit, was purchased from B.E. Taylor Company, a commercial and industrial real estate company aiming to establish an industrial park. Peninsular emerged as a prime property in this new industrial landscape. Over three thousand individuals attended the groundbreaking ceremony. Construction promptly commenced, and by December, they had completed a two-story administration building and a one-story factory spanning over 200,000 square feet. This new space could accommodate up to one thousand workers, adhering to the scheduled timeline.

The new planned layout for the factory floor, according to the *Michigan Manufacturer & Financial Record*, would make Peninsular "the most modern continuous molding plant in the world." The contrast between the old and the new ways of stove manufacturing was profound: "If an old stovemaker was taken into the mammoth new plant of the Peninsular Stove Co. in northwestern Detroit, he would find few earmarks of his old craft. The giant molders who once trundled molten iron around the foundry floor have been supplanted by two giants, electricity and compressed air, which never complain about overtime and do a great deal of the work that the old molders used to do, without having so much as a lame back to show for it."

On May 6, 1932, the Detroit business community was stunned when the Peninsular Stove Company filed for bankruptcy in U.S. District Court. Peninsular had prioritized modernizing its facilities over updating its product offerings, and unfortunately, it continued producing various outdated heating units that failed to attract buyers in a market shifting toward electrified central heating and other technological advancements. Moreover, retaining a skilled workforce at a time when the auto industry was paying workers top-dollar wages became an ongoing challenge, even predating the relocation to Brightmoor.

Sinking into significant debt, the company found no white knights willing to merge or acquire the struggling forty-five-year-old stove manufacturer. As a result, in June 1934, the Peninsular Stove Company was compelled to sell its Brightmoor facility to the Detroit Gasket Company—a supplier to the automobile and marine industries—as part of the bankruptcy liquidation process aimed at meeting creditor demands.

Chapter 7

MARKETING DETROIT'S CAST-IRON STOVES IN DETROIT AND AROUND THE WORLD

In the fiercely competitive market of durable goods, specifically cast-iron stoves, manufacturers in the Detroit area were aware that relying solely on the local market would eventually plateau in terms of growth. To expand and ensure success, they needed to broaden their market reach. Observing the strategies of their successful East Coast competitors like Vermont Castings and Round Oak Stove Company, as well as numerous other stove manufacturers nationwide, the Detroit-area manufacturers recognized that relying on middlemen could be detrimental to their stove-manufacturing endeavors.

Eliminating the Middleman

These middlemen, including wholesalers, retailers and catalogue/mail-order companies, played distinct roles in the distribution chain. Wholesalers bought goods in bulk from manufacturers and supplied them to merchants or smaller wholesalers. Retailers, such as hardware or specialty stores, exhibited these products for direct sale to consumers, while catalogue and mail-order companies advertised and facilitated sales through mail orders.

However, the use of middlemen proved less effective for products requiring distinct branding and product differentiation, such as cast-iron stoves. The uniqueness and craftsmanship of each brand were pivotal in influencing consumer choices within this competitive market.

Forward Integration

Recognizing this, Detroit-based manufacturers like Detroit Stove Works, Michigan Stove Company, Peninsular Stove Company, Art Stove Company and Vapor Stove Company adopted a strategy known as forward integration. This strategy involved these companies extending their control over distribution and sales, ensuring direct customer interactions and reinforcing brand recognition.

For instance, Michigan Stove Company established regional showrooms and warehouses in places like Chicago and Buffalo, enabling direct oversight of product distribution and promotion. While this approach demanded substantial investments to set up sales networks, including overseas, it fostered a more sustainable and profitable business model. Engaging directly with customers through their retail platforms allowed these manufacturers to gather valuable feedback, understand preferences and tailor their offerings accordingly.

Though backward integration, which involves controlling raw material sources and transportation, seemed appealing for these local stove manufacturers, the financial burden of investing in iron mines or owning Great Lakes iron ore carriers was too immense to seriously contemplate.

Central to forward integration was the creation of strong brand identities and consumer recognition through various marketing strategies like newspaper advertisements. Each Detroit stove manufacturer established unique brand names under which their products were marketed. For example, Detroit Stove Works marketed the Jewel line, Michigan Stove Company offered the Garland line, Peninsular presented the Sunburst and Sylvan lines, Art Stove showcased the Laurel line and Vapor introduced the Blue Star line.

Marketing with Trade Cards and Booklets

Marketing tools such as trade cards, akin to collectible baseball cards, were employed to promote cast-iron stoves. These colorful cards, distributed at

Left: Jewel stove trading card (front). *Author's collection.*

Below: Garland stove trading card (*left*: front; *right*: back). *Author's collection.*

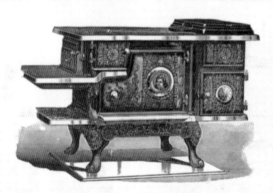

This page: Peninsular stove trading card (*top*: front; *bottom*: back).

This page: 1892 Garland stove trading card (*top*: front; *left*: back). *Author's collection.*

points of sale or events, depicted detailed drawings of stoves, highlighting their features and designs.

Furthermore, the Design Patent Act of 1842 provided substantial legal protection for these brand-name manufacturers, solidifying their positions in the market. As Howell J. Harris explains:

> *After the passage of the 1842 Design Patents Act, which also required makers of patented good to mark them with the patent date, it became possible to gain seven years' protection for a stove's outward shape and decoration as well as, or instead of, the existing fourteen years granted for an "improvement." Stovemakers made this law very much their own: they were responsible for four-fifths of all design patents in the 1840s and two-thirds in the 1850s. Unlike many other everyday goods, it was easy to cast the maker's and/or the model name, as well as the fact of being covered by patent, prominently and indelibly onto the surface of the stove itself. In this way, the name and the patent mark became key features of the stove's appearance, a permanent advertisement, and a deterrent, however imperfect, to counterfeiting.*

Harris contends that the stove industry pioneered the creation and sale of the initial universally embraced consumer durable, enjoying legal protection tailored primarily to this sector.

Presently, one can discover archival remnants of trade cards and other marketing materials issued by Detroit stove manufacturers at the Detroit Historical Museum, the Burton Historical Collection of the Detroit Public Library, the William L. Clements Library at the University of Michigan and the Walter P. Reuther Library at Wayne State University. Moreover, a simple search across online platforms like eBay, Reddit and Etsy reveals a plethora of original trade cards offered for sale, often at reasonable prices, catering to collectors' interests.

Chapter 8

THE STORY OF THE GIANT GARLAND STOVE

The grand vision of the colossal Garland stove—even in its most elemental form—originated with George Harrison Barbour, the vice president of the Michigan Stove Company. Collaborating with Jeremiah Dwyer, the company's president and founder, Barbour entrusted the task of designing and building an oversized iteration of the popular Garland Cooking Stove to William Keep, the plant superintendent.

Design and Construction of the Big Stove

Keep's initial considerations revolved around determining the exact, ideal dimensions of the final stove replica and the choice of materials for its construction. To make a significant impact and ensure it would tower over the company's appliance models on display at the upcoming Columbian National Exposition of 1893 in Chicago, Keep and his design team produced a working drawing that would materialize into a giant Garland Cooking Stove replica, towering at twenty-five feet in height, thirty feet in length and twenty feet in width.

The next critical decision was the selection of construction materials. Cast iron was quickly ruled out due to its immense weight, making the transportation of the giant stove nearly impossible. Moreover, the challenges and potential hazards associated with sand-casting patterns of unprecedented dimensions with molten metal were considerable. The only viable material

Postcard of the giant Garland stove in front of the Michigan Stove Company. *Author's collection.*

known to the skilled craftsmen at the plant was wood, primarily sugar pine and oak. While wood was relatively lightweight in comparison, it would need to be assembled and disassembled with relative ease, as it had to be transported by railroad cars to Chicago. There, the pieces would be assembled in the allocated thirty-by-forty-foot space on the east side of the Manufacturers and Liberal Arts Building in Jackson Park.

The identity of the artisans responsible for the wood carving remains shrouded in history. According to the Detroit Historical Commission, "Some sources say it was carved by John Tabaczuk in the employ of Jungwirth and Co., though this is unlikely as Tabaczuk was a lad of 14 in Poland at the time." A story recounted by Edna Franz of Lansing to *Free Press* columnist Mark Beltaire suggested that her grandfather John Wolf played a role in its construction. There was, in fact, a sizeable team of carpenters and woodcarvers, led by the chief sculptor Declan Shannon, who worked on the project. Shannon, an Irish immigrant from Ballymacoda, County Cork, had settled in Corktown, where he married and supported his family through his earnings from the Michigan Stove Company. Originally a carpenter, he had developed into a skilled "stairbuilder," a trade he inherited from his grandfather in Ballymacoda. His woodworking expertise proved invaluable, especially during the renovation of Holy Trinity Church, where his refined woodcarving skills were highly appreciated by churchgoers.

Upon completion of all the components for the colossal stove replica, including the elegantly curved legs and intricately carved apron, the skilled workmen meticulously assembled and securely bolted together the entire oven. The forthcoming task for the skeletal crew accompanying this mammoth stove to Chicago was to undertake the careful reassembly of its various parts, requiring three railroad cars for transportation.

Now weighing in at fifteen tons, the wooden structure demanded multiple layers of paint to transform into a convincing cast-iron stove. The legs, skirt and oven door handle were coated in a lustrous metallic silver, while the primary oven received a coating of flat black oil-based paint. This particular paint, comprising turpentine, a drying agent and pigmented oil, possessed excellent coverage, ease of application, adhesive properties and durability against the elements. Placards proudly bearing the Garland brand name adorned both ends, and a brand medallion took center stage on the stove's side.

Upon the paint's thorough drying and removal of all scaffolding, a diverse group of artisans and officials, including woodcarvers, carpenters, designers, molders, polishers and even company officials, along with both youthful apprentices and middle-aged men appearing older than their years, could only marvel at the breathtaking result. George Barbour's initial vision had evolved into something far beyond expectations—the giant Garland Cooking Stove would become a symbol of this industrial era.

The master sculptor, Declan Shannon, opted to forgo the journey to Chicago. Exhausted yet filled with pride and humility over his team's achievements, Shannon chose to remain at home, relishing precious moments with his wife, son and three daughters. Moreover, the seasoned traveling workmen were well versed in the assembly process, anticipating the opportunity to explore the World's Fair before returning to Detroit. Another team would be dispatched six months later, following the conclusion of the Exposition, to disassemble and transport the stove back to Detroit.

Success in Chicago

In Chicago, the colossal Garland stove lived up to its anticipated success. Within the bustling Manufacturers and Liberal Arts Building, a prominent thematic pavilion drawing a continuous flow of visitors, Matt R. Bigham, representing the Michigan Stove Company, enthusiastically welcomed every individual or small group. His well-practiced spiel, as reported by a Chicago newspaper, unfolded as follows:

> She holds the heavy-weight record in the stove line hard enough. She's 25 feet high, 30 feet long, and 20 feet wide. She weighs 20 tons [sic] and it took 3 [railroad] cars to bring her from our works in Detroit. Roughly speaking, she's as big as a country schoolhouse, and forty children could sit down in the oven. She's the same model as one of our Garland Cooking Stoves, and is the queen bee of the whole collection, which, you see, is extensive, and covers everything needful in the line of ranges and heaters. Anyhow, our exhibit cost over $25,000, and it takes a lot of stoves to be worth that much.

One might anticipate that the local news coverage of the giant Garland would be positive, and the *Detroit Free Press* review did not disappoint:

> To the World's Fair visitors, there are many attractions that do honor to Detroit and Michigan but possibly there are none more interesting from an industrial and commercial aspect than the exhibit of the Michigan Stove Company, of this city. Besides a very fine assortment of their famous Garland stoves, the company has on exhibition a mammoth stove which holds the attention of every visitor to the Manufacturers and Liberal Arts Building.

Local Chicago news coverage varied from straightforward flattery to outside rhapsodic in their portrayal. The *Chicago Daily Inter Ocean* exhibited a touch more enthusiasm: "To say that thousands of people daily survey the shiny sides of the black and silver leviathan is putting it mildly, and the Michigan Stove Works can congratulate themselves on having furnished one of the substantial and striking features of the Fair."

The *Chicago Tribune* continued to heap praise on the Michigan Stove Company's exhibit: "A varied and magnificent display of artistically designed stoves covers the space, and overall the great Colossus towers in its giant proportions, with one foot resting on each of the four corners of the exhibit."

Another writer for the *Chicago Daily Inter Ocean* waxed poetic: "'As a hen gathereth her chickens under her wings' might be paraphrased to read 'As a stove gathereth her stovelettes about her legs' and then fitly applied to the gigantic Garland which the Michigan Stove Company has enthroned as the Aegis of its exhibit on the east side of the Manufacturers building at the Exposition."

Few relics from the summer of 1893 would be as frequently and vividly recalled as the giant stove. The only true contender for attention would have been the 127-ton gun from Essen, Germany, showcased at the Krupp Pavilion. Known as the largest cannon in the world, the big gun was capable of firing a one-ton projectile over a distance of thirteen miles. Alternately, there was the Great Wheel erected at the heart of the Midway, designed by George Washington Gale Ferris. This "Ferris" Wheel boasted thirty-six pendulum cars, accommodating 40 passengers each. A single rotation of the wheel elevated 140 people to a height of 250 feet in the air, "giving to each passenger a magnificent view and a sensation of elevation akin to that of a balloon ascent."

THE BIG STOVE RETURNS HOME TO DETROIT

As planned, at the conclusion of Chicago's Columbian Exposition, a team of workers from the Michigan Stove Company returned to Chicago to disassemble and ship the giant Garland stove by rail. The decision to return the stove to Detroit immediately was a wise decision, for if they had decided to leave the stove at the Manufacturers and Liberal Arts Building until spring, for better weather, there may have been no giant Garland stove to return. On January 8, 1894, the building became a makeshift homeless encampment where a careless fire got out of control, and the entire building burned to the ground.

During the Exposition, a concrete pedestal base was being prepared in front of the Michigan Stove Works at Jefferson Avenue and Adair Street. Upon the return of the Garland, workers reassembled it atop the pedestal base. There, for more than a quarter century, the giant Garland stove would sit proudly for passersby and tourists and instill pride in the city's continued position of stove capital of the world.

In 1902, a bronze tablet measuring four by six feet was erected just beyond the cast-iron enclosure of the plant. This memorialized the Battle of Bloody Run and was positioned directly in front of the imposing Garland stove. The historical significance of this location dates back to 1763, when Chief Pontiac orchestrated a successful ambush against the British, resulting in substantial casualties on the very ground where the plant stood.

Over the following two decades, the Michigan Stove Company and the Detroit Stove Works thrived in proximity, a mere few miles apart. The escalating costs of union labor and the presence of redundant facilities prompted occasional familial discussions between the Dwyers and the Barbours. Unsurprisingly, these conversations inevitably gravitated toward the prospect of a merger.

In 1925, serious discussions unfolded among key figures from both companies, primarily involving the Dwyers and Barbours. By autumn, a preliminary agreement had been reached, solidifying in early December. The two companies, already intertwined by family bonds, would now coalesce under the unified banner of the Detroit-Michigan Stove Company.

As part of the consolidation process, employees from the Michigan Stove Company relocated to the Detroit Stove Works facility on the outskirts of Belle Isle. Meanwhile, the imposing Garland stove stood temporarily anchored at its original position on Jefferson and Adair. William T. Barbour assumed the role of president for the newly merged entity. In preparation for its eventual placement at the corporate headquarters, Barbour took custody of the bronze Bloody Run tablet to ensure its safety.

While the management busied themselves with the reorganization of the amalgamated facilities, the colossal Garland stove, reminiscent of an orphan, stood alone on the vacated Michigan Stove Company property for the ensuing two years. Despite the demands of the restructuring process, thoughts of the iconic stove were never far from their minds.

Move to New Home

In July 1927, William T. Barber enlisted the services of rigger Thomas Goodfellow to relocate a colossal stove to its new home. Goodfellow skillfully placed skids beneath the massive structure, securely fastening it with straps. The crane, under Goodfellow's control, smoothly lifted the unwieldy stove onto a flatbed truck, initiating its two-mile journey to what was hoped to be its permanent residence near the west entrance of Belle Isle.

Upon delicate placement of the stove's legs onto four concrete pedestals, it became evident that a fresh coat of paint was imperative to rejuvenate this iconic piece of early Detroit industry. Upon completion, the Garland resembled its appearance at the 1893 Chicago Columbian Exposition. A modest installation ceremony was scheduled for the subsequent week.

After ninety-one years under the stewardship of the Dwyer and Barbour families, among others, the Detroit-Michigan Stove Company officially entered history on April 26, 1955. Following a merger with Welbilt of Long Island, New York, the giant Garland stove fell under the ownership of the new corporate entity. Although Welbilt signage was affixed to the stove, a sense of its status as a historical landmark never really attached itself to the new company. Sadly, a reckoning almost came two years later. Welbilt chose to close its Detroit facility on June 30, 1951, leaving the industrial symbol unguarded and vulnerable to vandals. In the early morning hours of Sunday, November 23, 1950, a small, isolated one-alarm fire escalated quickly to a five-alarm fire, nearly obliterating one of the city's historical industrial landmarks. Firefighters focused on containing the blaze to the main building, deploying fifty pieces of equipment and involving 150 firefighters in a two-and-a-half-hour battle.

Rather than dismantling the stove during facility cleanup, Welbilt accepted an offer by Jack Shafer of Shafer Bakeries to lease the giant stove to promote his baked goods. He even envisioned constructing a giant loaf of bread to sit next to the big stove, though that particular enhancement never materialized.

Ten years later, Uniroyal Company purchased the property with the intent of using the space for parking and offered to donate the stove to the city. The fate of the stove now remained uncertain. Mary Dwyer Howarn, grandniece of Jeremiah Dwyer, expressed profound dismay at the stove's condition. In a letter to the editor appearing in the *Detroit Free Press* on December 7, 1964, Howarn described what the stove meant to generations of Detroiters:

> *There was a time when Detroit residents proudly escorted their out-of-town visitors to see the "Big Stove" which was ensconced majestically on the south side of East Jefferson Avenue between Adair and Leib Streets.*
>
> *It was included on all official sight-seeing tours and appeared on the luridly colored picture post cards of that era.*

After recounting the history of the stove from its inception, Howarn described its present situation at the southeast approach to Belle Isle Bridge:

> *It's still there—but a sorry sight! The once well-tended lawn surrounding it is now a weedy, sandy plot; the stove has been easy prey for vandals and the ravages of the elements—and (disgrace of all disgraces) it bears the taint of commercialism by being adorned with garish advertising.*

Couldn't this beloved landmark—a bit of "old Detroit"—be moved to a more suitable location, freed of its advertising burden, restored and preserved in keeping with its former importance?

Just moving it across the bridge approach to Gabriel Richard Park would seem to be the most practical and least costly plan. There it would really belong to the City of Detroit.

Big Stove, Big Move

Howarn's disappointment may seem to have gone unnoticed, but the sentiment resonated with many Detroiters who cherished the novelty and historical significance of the landmark. A dedicated group known as the Save the Stove Committee took action, locating a moving and cartage company that was willing and able to move the landmark stove to Woodward and Eight Mile Road with the approval of state fair officials. The stove was placed on concrete pillars within the city bus line's turning circle.

Transporting the Big Stove to the fairgrounds in 1965 presented a far greater challenge than its previous relocation just two miles down Jefferson Avenue in 1925. Royce Richards, the owner of Don Cartage Moving Company, generously agreed to move the massive and unwieldy Big Stove at no cost. However, his primary concern wasn't the absence of compensation for his time and the use of expensive equipment; rather, it was the risk of damaging this seventy-two-year-old landmark. A preliminary assessment of potential routes to the fairgrounds revealed that the stove's height posed a problem, as it wouldn't pass through viaducts and power lines along the way. Consequently, Royce had to chart a new course that bypassed these obstacles. Collaborating closely with his union workers and local utility companies, Royce meticulously devised a twenty-five-mile route that skillfully navigated around trouble spots and maneuvered under raised power lines.

The move was meticulously scheduled for 4:00 a.m. on a Saturday to minimize disruption to regular traffic. The truck hauling the trailer carrying the stove cautiously maintained a speed limit of no more than five miles per hour throughout the journey's numerous twists and turns. Despite meticulous planning, unforeseen issues arose, notably when early-morning church attendees disregarded "No Parking" signs along Woodward Avenue to attend services, obstructing the path. To negotiate a corner turn with closely parked cars, Royce ingeniously used a crane to elevate the trailer's back and swing it around the corner. After a grueling nine-hour journey, the Big Stove finally reached its new home at the fairgrounds. The stove was placed on concrete pillars within the city's bus line's turning circle. And here the Big Stove would remain for nearly a decade.

However, an unforeseen issue emerged. Mark Beltaire, a prominent voice in the Save the Stove Committee and a columnist for the *Free Press*, candidly admitted, "The one thing we didn't think of was maintenance. The stove appeared to be indestructible, but after 80 years of exposure to Detroit's climate, it was crumbling with dry rot."

In 1974, the ownership of the stove was subsequently transferred to the Detroit Historical Museum, which raised $5,500 for the stove's restoration and its permanent relocation near the riverfront, its original setting. Most of the funds raised went to JJR, an Ann Arbor–based architectural firm, for the assessment of the stove's current condition and its viability for restoration. The twenty-eight-page feasibility study concluded that the stove was structurally sound, with elements in "advanced stages of deterioration" deemed replaceable. However, the study strongly recommended identifying a temporary shelter before initiating any restoration efforts.

Into Storage at Detroit Historical Museum Warehouse

Regrettably, the proposed restoration came with a substantial price tag, ranging from $51,000 to $105,000. The final cost hinged on the level of generosity exhibited by the city and utility companies in contributing to the expenses associated with relocating stoplights, telephone cables and power lines. In light of these financial considerations, the historical museum made the decision to preserve the stove in its current state. To facilitate this, the stove was disassembled and relocated to storage at Historic Fort Wayne.

Under different circumstances, the tale of the Big Stove might have concluded unceremoniously: fragmented and forgotten in a warehouse near the Detroit River. For over twenty years, the memory had faded among many Detroit residents, while younger generations remained oblivious to this nostalgic slice of the city's history.

However, in an unexpected turn of events in 1993, Republican governor John Engler made a surprising move by appointing Democrat John Hertel, then chair of the Macomb County Board of Commissioners, as the general manager of the Michigan State Fair. This decision marked the fair's final opportunity to make a significant impact at its location on Woodward and Eight Mile Road, following a decline in attendance and growing public apprehension about safety concerns. Hertel was given a clear directive and a five-year timeline: revitalize the fair, make it financially viable or shut it down permanently.

The choice of a five-year span was deliberate, coinciding with the fair's 150th anniversary in 1998. This milestone offered Hertel a chance to help Detroit reclaim its robust agricultural and industrial legacy, while providing innovative family entertainment in the time-honored tradition of state fairs.

Despite a three-year slump in attendance, worsened by adverse weather conditions, the prospects for success seemed bleak. Yet, during a rainy afternoon, while tending to paperwork in his office, Hertel stumbled upon a small tin box left by someone months prior. Among various business cards inside, one stood out—a plain card with just the words "the stove" and a phone number. He dialed the number, connecting with a woman from the Detroit Historical Museum's warehouse.

"Do you have the giant stove there?" he inquired.

"We do, but unfortunately, it's in pieces," came the reply.

Restoration and Return to the Michigan State Fair

Responding swiftly, Hertel dispatched eight pickup trucks to collect the fragments of the Big Stove and transport them to an empty building on the fairgrounds. The painted wooden pieces, crafted in the 1890s, were spread out on the floor. The challenge ahead was twofold: could it be restored, and what would the restoration cost entail?

Hertel recognized a rare chance to revive a Detroit icon at a time when the city had lost many landmarks to progress's demolition. The Big Stove stood out as a distinctive piece of Detroit's heritage, evoking cherished childhood memories for numerous people. Beyond nostalgia, it represented a bygone era, linking Detroit to a time before automobiles, embodying the city's industrial history. It was deemed invaluable and worthy of preservation and restoration.

To kickstart the restoration, a fundraising campaign was launched, despite initial estimates of the daunting $300,000 cost. Undeterred, Hertel took to Detroit's WJR radio station, passionately advocating for support to restore the Big Stove. His plea resonated with Detroiters who recognized the project's significance. The campaign managed to collect $238,000 in cash, along with $100,000 in labor and materials, with substantial contributions from corporations and unions that played a pivotal role in reaching the funding goal.

What once lay scattered on the floor of an empty building at the fairgrounds was now being meticulously reassembled using cutting-edge technology from General Motors Tech Center, combined with the dedicated efforts of union carpenters and millwrights. All of this was achieved under the guiding vision of John Hertel.

The restoration of the Big Stove was completed just two days before the opening of the Michigan State Fair on Tuesday, August 25, 1998. A day prior, a special 150th anniversary ceremony and a formal rededication ceremony for the stove drew the attendance of Governor Engler, numerous local and state dignitaries and a small crowd of state fair enthusiasts at Eight Mile and Woodward. Alongside the renewed stove, fairgoers enjoyed freshly paved parking lots, refurbished facilities and heightened security provided by the Detroit and Michigan State Police. Furthermore, the entertainment was exceptional, featuring nightly performances by headliners like the Beach Boys, Ray Charles, comedian Bill Cosby and rocker Alice Cooper.

THE TRAGIC END OF THE BIG STOVE

The restoration of the Big Stove was indeed cause for celebration, given the loss of so much of Detroit and Michigan history through the years. For the next decade, the care and maintenance would be attended to by the staff of the Michigan State Fair. But then, unexpectedly, in 2010, the fair lost its funding. The governor basically decided the fair was expendable along with other budget cuts. Hertel moved on to work for the SMART transit system, but his concern remained for the safety of the Big Stove, which so many people had worked so hard and so long to restore. It remained at serious risk if left unattended at the closed fairgrounds.

Hertel made a number of calls to places that he believed would be an appropriate landing spot for the Big Stove. One after another, his pleases were rejected, either on the basis that the Big Stove was too big for the facility or that ongoing maintenance would be prohibitive. The Big Stove would drive attendance, Hertel countered, but his argument fell on deaf ears. The stove would remain on the grounds of the shuttered state fair.

Tragically, on August 12, 2011, sometime before 9:00 p.m., the Big Stove was reportedly struck by lightning during the evening rainstorm and was largely incinerated. What fragments remained—legs, the handle off the front of the stove and random pieces of painted sugar pine from the apron—were carted off to be stored, yet again, at the warehouse of the Detroit Historical Museum.

Hertel, however, believes the Big Stove likely suffered a more sinister fate:

> *I have served as a public government official almost my whole professional life (28 years in various state and county elected offices and 26 years in various appointed offices, and even though I have always been a Democrat, I was appointed to office 4 times by Republicans including 2 Republican governors).*
>
> *I mention all of this political, public service experience because it should clearly indicate that I am no stranger to often seeing things done one way and then having them explained in a completely different way.*
>
> *Sometimes these are honest mistakes and sometimes they are purposely mis-explained (there's a political word). At any rate, I want to tell a different story than the one that was put out publicly and officially about the destruction/burning of the Giant Michigan Stove.*
>
> *I was the person who re-discovered its parts, who came up with the idea to restore it, who managed the project and who personally raised most of the PRIVATE money that funded the 1998 restoration of the 1893 historical remnant of the COLUMBIAN EXPOSITION and Michigan's world-famous stove industry.*
>
> *One of the most important standards we strived for in restoring the Stove was to adhere to doing it as authentically as possible to the way it was originally built. Throughout the project we consistently used a variety of experts and historic information to stick to making it as close to original as possible.*
>
> *Because of this major rule we ultimately were able to save and re-use over 60 percent of the original wood (which also was the largest single expense because these pieces were stripped of over 100 years of lead paint layers).*
>
> *At any rate, we were so successful in achieving a historically accurate restoration that the stove won several awards for historical restoration.*

Now I will admit to purposely making the decision to stray from accuracy by personally ordering one major change from the original.

Except for one year in Chicago, the intact stove had always been on display outside, 365 days a year in all weather. Made of wood with a flat top, this was one of the reasons the stove deteriorated and was taken down in the 1960s.

Since the stove was 25 feet high and no one would see the top and since it would again be on display outside at the Michigan State Fairgrounds year round, I ordered that an 8 inch thick rubber roof be attached to the entire roof before we had the ceremonial unveiling in 1998, the night before the 150th anniversary of the opening of the Michigan State Fair.

I will keep it simple. Lightning does not strike rubber. Something or somebody else is responsible.

No part of the explanation put out by the City of Detroit or the State of Michigan is accurate.

Chapter 9

COMMEMORATING CHIEF PONTIAC'S REBELLION

The Michigan Stove Company's Bicentennial Tablet

In preparation for the 200th anniversary of Detroit's founding by Antoine de la Mothe Cadillac, Mayor William Maybury took the initiative to establish a Bicentennial Committee. This committee was tasked with organizing a weeklong celebration set to take place during the week of July 21, 1901. Comprising local historians, esteemed politicians and influential business leaders, the committee proposed various ideas, one of which was to create permanent markers, such as tablets, to commemorate individuals, locations and events significant to the history of Detroit.

A pivotal meeting early in this planning process occurred on the evening of Tuesday, February 28, 1900, at the Detroit Museum of Art. Over two hundred guests were invited to contribute their insights to the committee. The meeting was presided over by Senator Thomas W. Palmer, who began by touching on the rich tradition of marking historical sites and events with tablets, emphasizing the practice's prevalence in Europe over the past 2,400 years.

Palmer swiftly steered the discussion to the core objective of the evening:

> *The real purpose of this meeting this evening, I take it, is to suggest what are the best and most desirable points in our city to be remembered and identified....The places that I can think of at present that are worthy of tablets are where the four corners of old Fort Pontchartrain stood, the location of Fort Shelby, the site of the old church of St. Anne, the dwelling house of the first Catholic priest, the place where Cadillac landed, and Bloody Run, which is really a thousand feet south of Jefferson Avenue.*

Fort Pontchartrain, 1701. *Walter P. Reuther Library, Archives of Labor and Urban Affairs, Wayne State University.*

Palmer's version of the location of Bloody Run likely surprised those in attendance. The prevailing belief was that Pontiac's tree, a notable witness to events of July 31, 1763, had stood much farther north than the location Palmer was now suggesting as the site of these significant historical events. This discrepancy caught many attendees off guard.

> *I am afraid I am responsible for deception that has crept in concerning Bloody Run, and the old tree where Pontiac fell. When I was a boy, my companions and I used to take our shotguns and shoot into the trees in their neighborhood. We often laughingly said that in a hundred years from then people would be digging out those bullets and taking them home, treasuring them as relics of Indian insurrections. Our remarks came true but sooner than we expected. People have done that very thing and so firm an impression that "Bloody Run" was on Jefferson Avenue, near the stove works, has been made that I think it will be hard to teach some people different. Therefore, I think I have a moral duty to perform. First to try to blot out the false impression I helped to create, and secondly to help erect monuments that will tell where the great battlefield really was.*

George T. Barbour, the vice president of the Michigan Stove Company and a member of the Bicentennial Committee, assumed the task of acquiring the historical tablet commemorating Bloody Run. He then ensured its placement at the precise location where this significant event occurred, or at least in proximity to it, as Senator Thomas Palmer noted. Remarkably, this endeavor took place right at the heart of Barbour's own workplace.

Over the next eighteen months, the focus shifted away from addressing the issue of historical markers to more immediate priorities. These included organizing a series of historical reenactments, with the most significant one being the re-creation of Antoine de la Mothe Cadillac's landing with his entourage. These reenactments often featured the active involvement of local French heritage organizations and individuals appointed by the French government. Simultaneously, numerous lectures, musical performances and educational initiatives were in development.

Throughout these preparations, Charles Barbour remained unwavering in his commitment on behalf of the Michigan Stove Company. He was dedicated to producing and installing a historical tablet near the location of Bloody Run on Detroit's near east side, never losing sight of this obligation.

At 3:00 p.m. on Thursday, July 31, 1902, nearly five hundred Detroiters gathered in front of the Michigan Stove Company on Jefferson Avenue to witness the dedication of a large bronze tablet commemorating the Battle of Bloody Run, which had taken place 139 years previous at or near the site of the tablet installation. Donated to the city as a gift from the Michigan Stove Company, the tablet of bronze, measuring four by five feet, was modeled after a design by E.T. Schoonmaker and cast by the Chicago Bronze Company. The tablet is supported on either side by a colonial column, each surmounted by a wreath of laurel. Above the tablet, on an elegant shield, is represented the bridge and tree. On each side of the shield are shown the military swords and Native American tomahawks, the whole capped by the English crown.

The wording on the tablet proper was as follows:

> *This tablet marks the course of the historic stream called "Parent's Creek" after the battle of July 31, 1763, which took place near by. It was known as "Bloody Run." That battle closely followed the Indian outbreak known as the Pontiac Conspiracy and resulted in a loss to the English of fifty-six killed and wounded and the death of Captain Dalyell, of the British army. An old monarch of the forest, known as the Pontiac tree, stood in this vicinity until 1886 and was said to have been a silent witness of the combat.*

Below, on another shield, are the words, "July 31st, 1763. July 31st, 1902." And on the plinth at the base: "Erected by the Michigan Stove Co."

Bicentennial tablet. *Burton Historical Collection, Detroit Public Library.*

The dedication took place under the auspices of the Society of Colonial Wars, with the Society of the Sons of the American Revolution as invited guests. The Honorable Thomas W. Palmer, ex-president of the Society of Sons of the American Revolution, was expected to preside over the dedication but at the last minute was indisposed, and so Detroit mayor William C. Maybury presided instead. Theodore H. Eaton, governor of the Society of Colonial Wars, delivered a brief address, explaining the significance of the flags used by the society. The bronze tablet, which he later unveiled, was covered by a large British flag, signifying that the ongoing war was fought under the emblem of Great Britain. After the unveiling, George H. Barbour formally presented the tablet, on behalf of the Michigan Stove Company, to the City of Detroit, with Mayor Maybury accepting it on behalf of the people. Then followed an address by Professor A.H. Griffith, who related some of the events of the bloody conflict. The dedication was closed by the singing of "America."

Historical Background

In the aftermath of the French and Indian War (1754–63), the British Empire found itself in control of a vast territory in North America, including the Great Lakes region and the Ohio Valley. While this territorial expansion was seen as a victory for the British, it also sowed the seeds of discontent among the Native American tribes who inhabited these lands. One of the most significant conflicts to arise from this discontent was Pontiac's Rebellion, a violent and coordinated resistance movement led by Chief Pontiac of the Ottawa tribe. One of the pivotal battles was the Battle of Bloody Run at what is today Elmwood Cemetery on Detroit's near east side.

Origins of Pontiac's Rebellion

The origins of Pontiac's Rebellion can be traced back to the Treaty of Paris in 1763, which officially ended the French and Indian War. With the treaty, the British acquired vast territories previously controlled by the French, including the area around the Great Lakes and the Ohio River Valley. This change in colonial rule deeply affected the Native American tribes of the region, as they had established complex relationships and alliances with both the French and the British. Pontiac had expected relations, particularly with trade, to continue as they had with the French, but the British notion that they owned the land by defeating the French changed Pontiac's perspective on white encroachment.

The Native American tribes were now deeply distrustful of the British, fearing that their expansion would bring about an end to the profitable fur trade and encroach upon their ancestral lands. Additionally, the British imposed new policies that alienated and angered the Native Americans. The most significant of these policies was the Royal Proclamation of 1763, which sought to reserve lands west of the Appalachian Mountains to Native Americans and existing French settlers and whoever others might have settled the area. The British were wary of attempts by American colonists to leave the thirteen colonies to travel and settle farther west, making the situation economically and militarily unmanageable for the Crown. American colonists rebelled against such strictures and took scores of wagons westward toward the Ohio Valley. As more settlers from the East continued to cross the Appalachian divide, Native Americans came

Two men reading the bicentennial tablet in front of the giant Garland stove at the Michigan Stove factory headquarters. *Library of Congress.*

to view the British and their proclamation as further evidence of British treachery and deceit.

Particularly offensive and contrary to Native American interests were the attitudes and actions taken by Major General Jeffrey Amherst, who was recently knighted following his capture of Montreal, Quebec, which effectively sealed the fate of the French in the French and Indian War. Amherst, now placed in charge of the Great Lakes region, instituted polices of his own making that were squarely at odds with ones instituted by the French, who generally treated Native Americans with respect, cultivating both friendship and trade. The British in general, and Amherst in particular, treated Native Americans as less than human and certainly as a conquered people. Among Amherst's new policies, designed to weaken them economically and diminish their threat capacity, was limiting access to gunpowder and ammunition. Amherst also considered giving the Native Americans blankets infected with smallpox in order to wipe out as many of the "savages" as possible. No record exists to suggest whether Amherst followed through with such a plan.

Tensions Escalate

As tensions simmer, Chief Pontic emerged as a charismatic and influential leader among the Native American tribes of the Great Lakes region. Pontiac, an Ottawa chief, was determined to resist British encroachment and protect the interests of his people. In the spring of 1763, he began to rally various tribes, including the Ottawa, Ojibwa and Potawatomi, to form a united front against the British. The rebellion began with a series of attacks on British forts and settlements in the Ohio Valley and the Great Lakes region. One of the most notable early successes for Pontiac's forces was the capture of Fort Michilimackinac in northern Michigan by Ojibwa warriors from villages on Mackinac Island and along the Cheboygan River. These attacks were characterized by surprise and stealth, catching the British off guard.

The Battle of Bloody Run

One of the pivotal moments in Pontiac's Rebellion was the Battle of Bloody Run, which took place on July 31, 1763, near Fort Detroit. The British garrison inside the fort, led by Major Henry Gladwin, was under siege by Pontiac's forces. Insite the fort, the situation was growing dire as supplies dwindled and tensions ran high.

Desperate to break the siege, a group of British soldiers received information that the Native Americans were planning a surreptitious attack on the fort. Believing that a preemptive strike was their only chance at survival, the British soldiers decided to launch an attack on the Native American encampment, which was located near a bridge over Parent's Creek.

Unbeknownst to the British, the Native Americans had also received intelligence about their plan. As the British soldiers neared the bridge, Pontiac's warriors ambushed them in a brutal and bloody confrontation. The Battle of Bloody Run lived up to its name, as the creek and its banks became stained with the blood of both British and Native American fighters.

The Battle of Bloody Run was a ferocious and costly engagement for both sides. The casualty report paints a grim picture of the violence and destruction that unfolded that day:

- British forces, in their ill-fated attempt to launch a surprise attack, suffered significant casualties. Over sixty British soldiers

Chief Pontiac's Siege of Detroit. Painting by Frederick Remington. *Wikimedia Commons.*

were killed, and many more were wounded. The ferocity of the Native American resistance took the British by surprise, and they paid a heavy price for their audacious maneuver.
- On the Native American side, casualties were also substantial. While exact numbers are difficult to ascertain, it is estimated that around twenty to thirty warriors lost their lives in the battle. The Native Americans, however, maintained the advantage of local knowledge and defensive positions, allowing them to inflict heavy losses on the British.

The Battle of Bloody Run did not decisively tip the scales in favor of either side, but it did underscore the intensity and brutality of Pontiac's Rebellion. Despite the heavy losses on both sides, the siege of Fort Detroit continued for several months, with neither the British nor the Native Americans able to secure a clear victory.

In the larger context of Pontiac's Rebellion, the conflict continued for several years, with sporadic violence and diplomacy. Ultimately, Pontiac himself made peace with the British in 1766, marking the end of his active role in the rebellion. The rebellion, however, left a lasting impact on the relationship between Native American tribes and the British Empire in

North America, leading to significant changes in British policies and their treatment of Native Americans in the region.

The Battle of Bloody Run serves as a poignant reminder of the complex and often tragic history of colonial expansion in North America and the resistance it sparked among the Indigenous peoples of the continent. It stands as a testament to the courage and determination of Chief Pontiac and the Native American tribes who fought to protect their way of life in the face of overwhelming odds.

The Disappearance of the Bloody Run Tablet

For the next twenty-five years after its installment, the Bloody Run tablet remained attached to the cast-iron fence that ran parallel to Jefferson Avenue in front of the Michigan Stove Company and in the shadow of the giant Garland stove that sat on a concrete pedestal inside the fence. Untouched by fire in 1907 that nearly felled the giant Garland stove, the historic tablet remained on site. Upon the merger of the Detroit Stove Works and the Michigan Stove Company in 1926, William T. Barbour assumed possession of the tablet as the new company prepared to move farther east, near the entrance of Belle Isle.

The whereabouts of the historic tablet is unknown. Perhaps following the merger of Detroit Stove Works and the Michigan Stove Company it was installed at the same time the Big Stove was moved to its new location. Perhaps after the merger with Welbilt and its subsequent plant closure, the tablet was stolen, vandalized, damaged or destroyed during the fire at the abandoned plant.

What happened to the Bloody Run tablet remains a "history mystery," but perhaps someone has information that would result in the return of the tablet to the city to whom it was dedicated in 1902 or at least to provide closure concerning its fate.

CONCLUSION

Detroit's historical narrative extends well beyond the advent of the moving assembly line and the rise of the automobile industry. However, a cursory glance at the local library shelves' books and journals dedicated to Detroit might erroneously suggest otherwise. The eighteenth and nineteenth centuries in Detroit's history brim with captivating tales featuring larger-than-life figures and significant strides in agriculture, commerce and industry. Notably, the latter half of the nineteenth century holds immense potential for deeper exploration, particularly following the discovery of iron ore in the Upper Peninsula and the establishment of the Soo Locks. These milestones warrant greater attention and investigation.

The irreparable loss of two significant historical relics—the Big Stove and the Bloody Run tablet—is profoundly disheartening, especially considering that their demise could have been easily prevented. Regardless of one's perspective on the Big Stove—whether it was regarded as a pivotal artifact representing late nineteenth-century Detroit industry or simply viewed as a sentimental emblem of local charm—it undeniably warranted preservation from neglect. Initially transported back from Chicago to commemorate Detroit's role in American industry during the nation's 400th anniversary of Christopher Columbus's arrival in the New World, this relic held a prominent place in Detroit's narrative. From its original location at Jefferson and Adair to its subsequent shifts to the Belle Isle entrance ramp and, finally, its permanent residence at the state fairgrounds at Eight Mile and Woodward, the Big Stove consistently had custodians dedicated to its safeguarding—

Conclusion

until the ill-fated evening of August 12, 2011. Numerous local and regional institutions existed with the capability to rescue the Big Stove and potentially shield it from its tragic fate. Unfortunately, no proactive steps were taken, resulting in the loss of this invaluable artifact from nineteenth-century Detroit industry, consigning it to the annals of history, now only accessible through aged photographs and videos.

Similarly, the Bloody Run tablet, an early twentieth-century historical marker, holds significant historical importance as an artifact. Presented to Detroit by the Michigan Stove Company in 1902, this tablet commemorates the 1763 clash between Native American warriors and British troops at Parent's Creek bridge. The British attempted an assault on the Native American encampment, only to face an ambush orchestrated by Chief Pontiac's warriors, occurring on or near the land later owned by the Michigan Stove Company. This tablet, measuring two feet by four feet and made of bronze, bears immense historical value. While it's conceivable that it hasn't been stolen or destroyed, there remains the possibility that it has been misplaced or is currently in someone's possession. Plausibly, this heavy artifact might still reside within the city limits. Nonetheless, given its status as a gift to the city, it ideally belongs on public display, such as at the Detroit Historical Museum.

A deeper loss for those seeking a comprehensive grasp of Detroit's history lies in the absence of recognition for the places, industries and people that formed the bedrock of the city's social, commercial and industrial framework before the era of automobiles. Detroit's pre-twentieth-century prosperity was closely intertwined with its diverse industrial landscape, spanning various locations across the city. Key stove manufacturers such as Detroit Stove Works, Michigan Stove Company and Peninsular Stove Company strategically positioned themselves along the riverfront and close to rail lines. Specifically, Detroit Stove and Michigan Stove were situated east of Woodward Avenue along Jefferson Avenue, while Peninsular was west of Woodward on Fort Street, occupying the area that currently houses the main Post Office Distribution Center.

Southwest Detroit, encompassing Delray, Woodmere and Springwells, stood as a heavily industrialized hub. Before the automobile era, this region housed nearly a dozen major industries, each employing over five hundred workers. Among them, the Springwells complex of American Car and Foundry churned out freight cars while the Detroit Dry Dock Company, situated along the riverfront, constructed some of the Great Lakes' largest ships. Additionally, the area hosted more than twenty brickyards, staffed

Conclusion

by over seven hundred workers who collectively produced a staggering one hundred million bricks annually.

Contrasting the east side's presence of the pharmaceutical giant Parke-Davis, southwest Detroit boasted the Frederick Stearns Company, an originally situated pharmaceutical manufacturing entity at Marquette and 21st Street before relocating operations to East Jefferson Avenue.

Notably, comprehensive research has delved into the Michigan Alcali Company, slightly downstream in Wyandotte. Arthur Pound's *Salt of the Earth* (1940) meticulously traces Detroit's crucial subterranean salt reserves, vital for manufacturing salt derivatives like soda ash essential in plate glass production. Yet scant historical exploration exists regarding numerous pivotal metalworking shops integral to the early automotive industry, such as the Detroit Copper and Brass Rolling Mills, which eventually became a primary parts supplier for the burgeoning Ford Motor Company.

Information on late nineteenth-century commercial and industrial leaders remains scarce. While Mayor Hazen Pingree's significant political achievements are well documented in Melvin Holli's *Reform in Detroit* (1969), his prosperous stint as a businessman alongside Charles Smith forming the Pingree and Smith shoe company in 1866 receives minimal attention. Despite growing into one of the Midwest's largest shoe and boot manufacturers, producing over half a million footwear pieces and employing seven hundred workers, scant details persist about Pingree's entrepreneurial career.

Similarly unexplored is the legacy of Eber Brock Ward, Detroit's inaugural millionaire industrialist. Flourishing predominantly in shipbuilding, iron and steel manufacturing, Ward diversified his pursuits by establishing sawmills, venturing into silver mining in the Upper Peninsula and serving as the president of the Flint and Pere Marquette Railroad Company.

Numerous significant figures like James McMillan and John Stoughton Newberry remain uncharted in biographical narratives, despite their prolific involvement in joint commercial and industrial ventures like the Michigan Car Company and the Detroit Car Wheel Company, paralleled by successful careers in politics. Dexter Mason Ferry, another notable figure, founded D.M. Ferry & Company, an early seed-vending business retailing seeds in small packets, warranting further biographical exploration. Attention is also deserved on Thomas and Joseph Berry, proprietors of the world's largest varnish factory.

The population explosion that unfolded in Detroit during the fifty-year pre-auto era from 1850 to 1900 stands in stark contrast to the subsequent fifty years of the auto era from 1900 to 1950. At the onset of the automotive revolution, Detroit boasted a population of 21,019 in 1850, which surged to 285,704 by 1900, ranking it as the nineteenth-largest city in the nation and the fourth largest in the Midwest, trailing behind Chicago, Cleveland and Cincinnati.

The ensuing half century, especially post-1913 following Henry Ford's introduction of the first moving assembly line, witnessed an extraordinary surge in population propelled by the expansion of automobile manufacturing. By 1920, the influx of both Black and white migrants from the South, alongside immigrants from Europe and the near East, catapulted Detroit to the status of the fourth-largest city in the United States, boasting 993,678 inhabitants. Subsequent decades saw staggering growth: 1,568,662 residents in 1930, 1,623,452 in 1940 and a pinnacle of 1,849,568 by 1950.

Detroit's metamorphosis from the Stove Capital of the World to the Automobile Capital of the World marked an immense and intricate shift of an entirely different magnitude. This transformation, however, was significantly aided by a robust industrial foundation that directly and substantially fueled the rapid ascent of the automotive industry. Any comprehensive account of the city's history that overlooks or belittles the contribution of these early industries, such as the stove industry, fails to be truly comprehensive.

BIBLIOGRAPHY

Berkowski, Neala. "Detroit's Culture and Growth Shaped by Immigrant Communities." *Michigan Daily*, February 1, 2015. www.michigandaily.com/uncategorized/detroits-immigration.

Berry, Thomas. *The Book of Detroiters: A Biographical Dictionary of Leading Living Men of the City of Detroit*. Edited by Albert Nelson Marquis. Chicago: A.W. Marquis & Company, 1908, 57–59.

Bienkowski, Brian. "Uncovering Buried Creek Could Spur Detroit Development, Ease Sewer Woes." Great Lakes Echo, August 18, 2011. greatlakesecho.org/2011/08/18/detroit%e2%80%99s-long-buried-bloody-run-would-flow-again-through-planned-development.

Blum, Peter H. *Brewed in Detroit: Breweries and Beers Since 1830*. Detroit: Wayne State University, 1999.

Bock, Gordon. "The History of Old Stoves." Old Houses Online, March–April 2008. www.oldhouseonline.com/kitchen-and-baths-articles/kitchen-appliances/history-of-the-kitchen-stove.

Brewer, Priscilla J. *From Fireplace to Cookstove: Technology and the Domestic Ideal in America*. Syracuse, NY: Syracuse University Press, 2000.

Burton, Clarence M. *A Compendium of History and Biography of the City of Detroit and Wayne County, Michigan*. Chicago: H. Taylor & Co., 1909.

Chicago Daily Inter Ocean.

Chicago Tribune.

"Christian H. Buhl." In *Compendium of History and Biography of the City of Detroit and Wayne County, Michigan*, edited by Clarence Monroe Burton, 320. Chicago: Henry Taylor & Co., 1909.

Bibliography

Christian, Rudy R. "Resurrecting the Detroit Central Farmer's Market." *Journal of Traditional Building Architecture and Urbanism*, November 2022, 209–18. www.traditionalarchitecturejournal.com/index.php/home/article/view/595/125.

Cooper, Patricia A. *Once a Cigar Maker: Men, Women, and Work Culture in American Cigar Factories, 1900–1919.* Urbana: University of Illinois Press, 1987.

Detroit Free Press.

Detroit Historical Society. "Detroit Dry Dock Company." detroithistorical.org/learn/encyclopedia-of-detroit/detroit-dr-dock-company.

———. "Ferry, Dexter M." detroithistorical.org/learn/encyclopedia-of-detroit/ferry-dexter-m.

Detroit News.

Detroit Times.

Digital Research Library of Illinois History Journal. "Chief Pontiac, the Red Napoleon. Head of the Ottawa and Organizer of the First Indian Confederation, the Illinois." January 11, 2023. drloihjournal.blogspot.com/2023/01/pontiac-chief-of-ottawas-history.html.

Dixon, Michael M. *Motormen & Yachting: The Waterfront Heritage of the Automobile Industry.* Fairhaven, MI: Mervue Publications, 2005.

Dolan, Kerry A. "How to Blow $9 Billion: The Fallen Stroh Family." *Forbes*, July 20, 2014. www.forbes.com/sites/kerryadolan/2014/07/08/how-the-stroh-family-lost-the-largest-private-beer-fortune-in-the-u-s/?sh=635342603d13.

Dusty Diary. "Late 19th-Century Gasoline Stoves: Cooking on a Bomb Used to Be Normal." ypsiarchivesdustydiary.blogspot.com/2009/05/late-19th-century-gasoline-stoves.html.

Farmer, Silas. *History of Detroit and Wayne County and Early Michigan: A Chronological Cyclopedia of the Past and Present.* Detroit: Silas Farmer & Co., 1890.

George Jerome & Co. "History of George Jerome & Co., Consulting Civil Engineers and Surveyors." Georgejerome.com/history.

Harris, Howell John. "Conquering Winter: U.S. Consumers and the Cast-Iron Stove." *Building Research and Information* 36, no. 4: 337–50, DOI: 10.1080/09613210802117411.

———. "The U.S. Stove Industry, c. 1815–1875: Making and Selling the First Universal Consumer Durable." *The Business History Review* 82, no. 4 (Winter 2008): 701–33.

Holli, Melvin G. *Reform in Detroit: Hazen Pingree and Urban Politics.* New York: Oxford University Press, 1969.

Hyde, Charles K. "'Detroit the Dynamic': The Industrial History of Detroit from Cigars to Cars." *Michigan Historical Review* 27 (1): 57–73.

Ihnat, Natalie. "Working Conditions in the Iron Industry." Iron Lives. ironallentownpa.org/introduction/working-conditions-in-the-iron-industry.

Klug, Thomas. "The Deindustrialization of Detroit." In *Detroit 1967: Origins, Impacts, Legacies*, edited by Joel Stone, 67–75. Detroit: Wayne State University Press, 2017.

———. "A Look Back…The Early Years of the Employers' Association of Detroit: The Ideal Manufacturing Company's Long Battle with Organized Labor." June 5, 2023. ASE Online. www.aseonline.org/News-Events/ASE-News/EverythingPeople-This-Week/a-look-back-the-early-years-of-the-employers-association-of-detroit-the-ideal-manufacturing-companys-long-battle-with-organized-labor.

Lawton, C.A. "Everything You Need to Know about a Casting." calawton.com/metalcasting-dictionary.

Loynd, Harry J. *Parke-Davis: The Never-Ending Search for Better Medicines*. New York: Newcomen Society in North America, 1957.

McKinney, James P. *The Industrial Advantages of Detroit*. Detroit: James P. McKinney, 1890.

Mitchell, James J. *Detroit in History and Commerce*. Detroit: Rogers & Thorpe, 1891.

Nagle, Michael W. *The Forgotten Iron King of the Great Lakes: Eber Brock Ward, 1811–1875*. Detroit: Wayne State University Press, 2022.

National Parks Service. "Parke-Davis and Company. Pharmaceutical Plant." www.nps.gov/places/parke-davisandcompanypharmaceuticalplant.htm#:&:textOperating%20a%20small%20drug%20store,world's%20largest%20sellers%20of%20pharmaceuticals.

Palmer, Thomas Witherell. *Detroit in 1837: Recollections of Thomas W. Palmer*. Ann Arbor: University of Michigan Library, 1922.

Parke-Davis & Company. *Fifty Years of Manufacturing, Pharmacy and Biology: Jubilee Souvenir, 1866–1916*. Detroit: Parke-Davis & Company, 1916.

Parkins, Almon Ernest. *The Historical Geography of Detroit*. Lansing: Michigan Historical Commission, 1918.

Piazza, Gregory C. *The Sage of the Log Cabin: The Life of Senator Thomas W. Palmer*. Pittsburgh, PA: Dorrance Publishing, 2018.

Pound, Arthur. *Salt of the Earth: The Story of Captain J.B. Ford and Michigan Alkali Company, 1890–1940*. Boston: Atlantic Monthly, 1940.

Roney, Brendan. "Don't Forget the Stove City." November 16, 2011. detroithistorical.wordpress.com/2011/11/16/dont-forget-the-stove-city.

———. "Moving Day for the World's Largest Stove." detroithistorical.org/blog/2014-02-14-moving-day-world%E2%80%99s-largest-stove.

"The Settlement of Michigan." project.geo.msu.edu/geogmich/settlement_of_MI.html.

Stevenson, Chris. "How Much Firewood Did Colonial Americans Use?" chrisstevensonauthor.com/2016/01/24/how-much-firewood-did-colonial-americans-use.

Strasser, Susan. *Never Done: A History of American Housework*. New York: Pantheon Books, 1982.

Strobe, Ryan. "Selling Stoves: Trade Cards from Michigan Stove Companies." March 18, 2015. detroithistorical.wordpress.com/2015/03/18/advertising-at-the-turn-of-the-century-trade-cards-and-michigans-stove-compaies.

Taylor, Paul. *"Old Slow Town": Detroit During the Civil War*. Detroit: Wayne State University Press, 2013.

Tien Vu, Dinh. "Foundry Safety Precaution in Casting Workshop." vietnamcastiron/foundry-safety.

Tuttle, Charles Richard. *General History of the State of Michigan: With Biographical Sketches, Portrait Engravings, and Numerous Illustrations: A Compete History of the Peninsular State from Its Earliest Settlement to the Present Time*. N.p., 1873.

White, John H., Jr. *The American Railroad Freight Car: From the Wood-Car Era to the Coming of Steel*. Baltimore, MD: Johns Hopkins University Press, 1993.

Zunz, Olivier. *The Changing Face of Inequality: Urbanization, Industrial Development, and Immigrants in Detroit, 1880–1920*. Chicago: University of Chicago Press, 1982.

ABOUT THE AUTHOR

Gerald Van Dusen is a writer and historian with a special interest in Detroit and its contributions to the American story. He is the author of *Detroit's Birwood Wall*, winner of the 2020 Library of Michigan Notable Book Award; *Detroit's Sojourner Truth Housing Riot of 1942*; and four other books. A recipient of numerous awards for innovations in teaching, learning and technology, Van Dusen is a father of four and a grandfather of two and resides with Patricia, his wife of forty-five years, in Redford, Michigan.

Visit us at
www.historypress.com